Created and Directed by Hans Höfer

INSIGHT GUIDES

PRAGUE

Edited by Joachim Chwaszcza
Photography by Bodo Bondzio and Joachim Chwaszcza
Managing Editor: Tony Halliday
Editorial Director: Brian Bell

HOUGHTON MIFFLIN COMPANY

APA PUBLICATIONS

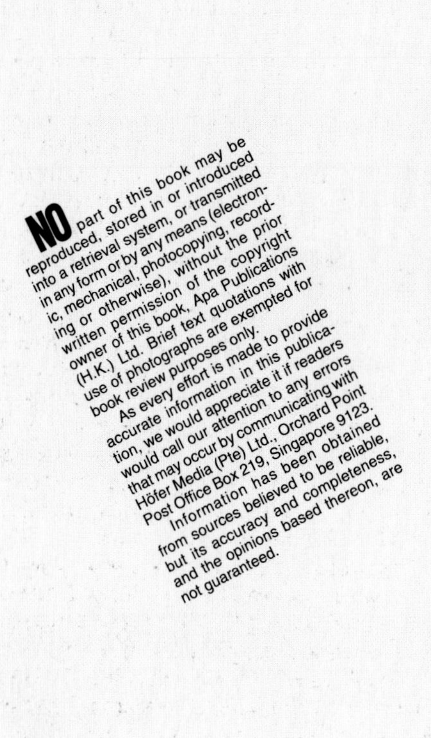

PRAGUE

Second Edition

Printed in Singapore by Höfer Press Pte. Ltd

Distributed in the United States by:
Houghton Mifflin Company
2 Park Street
Boston, Massachusetts 02108
ISBN: 0-395-66238-9

Distributed in Canada by:
Thomas Allen & Son
390 Steelcase Road East
Markham, Ontario L3R 1G2
ISBN: 0-395-66238-9

Distributed in the UK & Ireland by:
GeoCenter International UK Ltd
The Viables Center, Harrow Way
Basingstoke, Hampshire RG22 4BJ
ISBN: 9-62421-086-1

Worldwide distribution enquiries:
Höfer Communications Pte Ltd
38 Joo Koon Road
Singapore 2262
ISBN: 9-62421-086-1

ABOUT THIS BOOK

Prague today is one of Europe's most exciting cities, representing once again an unparalleled cross-section of all the grand themes of Western civilisation. Removed from the mainstream of tourism for four decades, the city proved irresistible to tens of thousands of Westerners as soon as the Iron Curtain lifted, and within weeks there was hardly a hotel room to be had. Demand is likely to overstretch facilities for years, so there's something to be said for going now rather than postponing a visit until demand has calmed down. One way in which this book, a completely revised and updated edition, has taken account of the changes is to provide a greatly enlarged Travel Tips section, giving much more information than before about things to see and do, and places to wine and dine.

There's certainly a lot to see. View Prague from the parapets of Hradčany Castle and it appears one of Europe's most fortunate cities: its skyline was never touched by the ravages of war and its essential countenance has not been greatly altered by the addition of modern eyesores. The historical centre, whose rooftops and spires reflect the golden patina of the midday sun, clings to the gently curving bend in the Vltava. Its banks seem to be only just held together by the delicate constructions of its bridges: on the one side, the Lesser Quarter; on the other, the Old Town.

Prague's architectural legacy not only provides testimony to the glories, uncertainties and tumults of bygone Bohemian days, but also reflects the development of an urban life extending back for more than 1,000 years; from its Romanesque basilicas, to its mighty Gothic churches, baroque palaces and right up to its magnificent boulevards of the "foundation years" laid out towards the end of the 19th century, with their wonderful art nouveau facades. After years of painstaking restoration, important architectural sights such as the Charles Bridge and the Old Town Square now bask once again in their former glory.

But Prague isn't just an architectural museum. The city has also been moulded by its cultural traditions and achievements. It has been associated with some of the great names of music and literature. Mozart's *Prague Symphony* and *Don Giovanni* were first performed in Prague, and the works of the great Czech composers Smetana, Dvořák and Janáček are still commemorated in the annual spring music festival. One of the city's many venerable taverns, the Chalice, provided the setting for the rebellious antics of Schwejk, immortalised by Jaroslav Hašek in *The Good Soldier Schwejk*. The enigmatic works of Franz Kafka are also inextricably linked with life in Prague.

The project editor assigned to this book was **Joachim Chwaszcza**, a Munich-based freelance photographer and writer. Although he is a regular Insight Guides contributor, this book had a special meaning for him because one of his parents came from Czechoslovakia. Chwaszcza also contributed several articles and pictures to the book.

The Authors

His sister, **Christine Chwaszcza**, who studied political science in Munich, compiled the first part of the historical introduction, and outlined developments in the city during the glorious reign of Charles IV.

Eva Meschede, who also studied political science in Munich, works as a freelance for

J. Chwaszcza

C. Chwaszcza

Meschede

Filip

Wagner

newspapers and radio. Here she describes events affecting the city from the rule of the Habsburgs to the annexation of Bohemia by Adolf Hitler in the 1930s.

Ota Filip looks at more recent developments. Born in Czechoslovakia and educated at the Academy for Journalism, he has lived in Munich since 1975, working as a freelance writer and author. He has written 10 novels, the latest of which is *Café Slavia.* His works have been translated into seven languages.

Vilem Wagner was born in Prague and lived there till he was 21. He studied the violin and music in Prague, Munich and Hamburg and has worked in films and television. Wagner lives in Hamburg, but he remains Czech in his heart and soul. He takes us on a tour of the historic castle of Hradčany, the Lesser Quarter and the Old Town, and then covers music and nightlife.

Johanna von Herzogenberg was born in Sichrow, Northern Bohemia. She studied German and Art History in Prague and Tübingen and graduated from Charles University in Prague. She has written for a number of art history publications and works for newspapers and radio. Her contributions here include tours of the Strahov monastery, the ancient Vyšehrad Castle and the Loreto shrine.

Jana Kubalova, director in the Glass and Ceramics section of Prague's Craft Museum, contributed a short history of Bohemian glass.

Dr Frantisek Kafka was a civil servant in the first post-war government from 1947 onward, and leader of the Jewish community in Prague from 1974 to 1978. He has had more than 20 books published. Here he writes about the *Golem*, that fearsome monster of medieval Prague.

Franz Peter Künzel describes the literary giants of the city. He was one of the first to brave the Cold War and editorial caution to popularise Czech literature in the West. He translated Hrbal and many other well-known authors and received the Translators' Prize from the Czech Writers' Union.

Marc Rehle is an architect who lectures in Design and Preservation of Historic Monuments at the Technische Universität in Munich. His interest in Eastern Europe has continued to grow since his student years, and he has explored Czechoslovakia and other East European countries during frequent visits as part of his ongoing search for important but forgotten 20th-century buildings.

The Photographers

Many of the photographs are the work of **Bodo Bondzio,** who is married to a native of Prague and is a regular Insight Guides contributor. Supplementary photos were taken by Joachim Chwaszcza and **Jens Schumann**. **Pavel Scheufler** provided several images from his rich archives of historic pictures – a collection accumulated over 20 years.

Special thanks go to **L.** and **P. Vilas**, **J. Kafka** and **L. Adamek** for their invaluable support. The original text was translated into English by **Susan James**.

Revisions and updates for this new edition, including major additions to the Travel Tips, were carried out by **Werner Jakobsmeier** in Munich. **Chris Pomery**, a British journalist based in Prague, provided the book with the latest information on political developments. This edition was produced in Insight Guides' London editorial office under the supervision of managing editor **Tony Halliday**.

von Herzogenberg

Kafka

Künzel

Rehle

Bondzio

History

Places

Features

Maps

TRAVEL TIPS

For detailed information
see page 211

AT THE HEART OF EUROPE

Going to Prague is fun again. Wherever you wander in the city, you can sense the new spirit. Of course, you're bound to encounter a certain amount of scepticism about the future: after all the decades of socialist mismanagement, the economic recovery hasn't happened as quickly as some hoped it might. But despite this, and despite the break-up of the Federal Republic into separate Czech and Slovak states, Prague, the Golden City, seems to be regaining its old confidence.

Life on the streets has changed. Since the momentous "velvet revolution" at the end of 1989, the arts, great and small, have once more found their place in the open. Theatre, exhibitions, debates – all the things that Prague residents used to long for – are now possible. The availability of Western news magazines was taken for granted within the first few weeks. Nowadays you walk through a city with all the normal signs of life in its streets.

Hordes of tourists are descending, all eager to see the changes for themselves. Yet despite this invasion, Prague essentially remains a Bohemian city. While its legendary days as a European centre of writers and artists have all but vanished, much of the former lifestyle, with its coffee-house ambience, has been preserved – in spite of Stalinism. But a melancholy, even morbid air suffuses the narrow streets. In the magical squares, with their continuous mystery plays of light, it is almost possible to imagine that Rabbi Löw's Golem, the monster that inspired Frankenstein, might still be lurking somewhere in the shadows.

Prague is not yet a fashion city, much less a modern one. Certainly, one can detect modern elements, and the new policies will doubtless bring about visible changes in the years to come. But the charm of the city doesn't rest in its elegant stores. It stems far more from the atmosphere imbued by the narrow alleyways of the Malá Strana, the baroque palaces and churches, and of course the Hradčany, the mighty castle complex that has dominated the city for almost 1,000 years.

It is Prague's great cultural tradition which has worked for centuries to mould this city and its people, and which continues to fascinate us. There is much which is still in poor condition and it will be years before Prague has completely recovered from the socialist era. Perfectly restored areas such as the Old Town Square can do as little to disguise this as the well-stocked stores on Wenceslas Square. But the first decisive steps have been taken. Prague has once again found its place at the heart of Europe.

Preceding pages: Prague *circa* 1860; young woman of Prague; Charles Bridge; dome of St Nicholas in the Malá Strana. Left, the Old Town Hall.

LIBUŠE AND THE BOHEMIAN KINGS

The founding of Prague is surrounded by legend. The chronicler Cosmas tells how Libuše, the wife of Přemysl, persuaded her husband to search out an unimpressive village on the banks of the Vltava and found a city there. She prophesied great things for the city: "The time will come when two golden olive trees will grow in this city. Their tops will reach the seventh heaven and they will shine throughout the world through signs and wonders."

According to the legend, Přemysl and his followers went to the place Libuše described and there founded Prague.

The rise of the Přemyslid Dynasty: Legend apart, archaeological finds confirm that the area around Prague has been inhabited since Neolithic times. However, the evolution of the city as the political and cultural centre of Bohemia is very closely bound up with the rise and the power of the Přemyslid dynasty. The battle for supremacy in the Bohemian and Moldavian regions between the Slavníkovci and the Přemyslids was won by the Přemyslids towards the end of the 9th century. Their rulers moved their residence to a strategic rocky outcrop on the right bank of the Vltava, and this is where they built their first castle.

Settlements developed in the area beneath the castle. At first these were inhabited only by people directly supplying the castle, but later craftsmen also settled here. In the early 10th century the Vyšehrad (originally known as the "Chrasten") was built, on the same bank of the Vltava, but some distance south of the old castle. The Hradčany was founded a little later on an equally commanding left-bank site some distance downstream. Under this defensive protection, the Přemyslids consolidated a political base centred on Prague that was to be the nucleus of the Bohemian state and that enabled the natural trade advantages of the settlement to develop.

One of the early princes of the dynasty was St Wenceslas, the patron saint of Czechoslovakia and the "Good King Wenceslas" of the Christmas carol. He put his duchy under the protection of King Henry the Fowler of Germany, and encouraged German missionaries to come to Bohemia, but was then murdered by his pagan brother Boleslav in about 929. Boleslav's rule (936–967) saw Bohemia consolidating itself against the German threat. But after the expulsion of the Magyars, it cultivated its connections to the West, extending them as far as Rome.

Prague quickly became an important centre for trade. In 965, the Jewish merchant and traveller Ibrahim ibn Yaqub wrote that "the city of Prague is built of stone and lime and is a busy trading centre". This might be a slight exaggeration, since the clay walls of the castle were probably not replaced with stone until the 12th century.

Founding of the Bishopric of Prague: A decisive point in the development of Prague was the success of Boleslav II in obtaining the consent, long withheld, of the Holy Roman Emperor for the founding of a bishopric in Prague in 973. The first bishop was the monk Adalbert, trained in the cathedral school in Magdeburg (an important centre of the Slav Mission) and appointed in 982. Adalbert fled from Prague on several occasions and died in Poland. His relics were seized by Břetislav I (1040–55) during a campaign in Poland and brought back to the city. The plan to use the relics to raise the status of the Prague cathedral came to nothing, but Prague greatly increased in international importance during Břetislav's reign.

Břetislav's successor Spytihnev II (1055–61) is credited with the expulsion of German merchants from Prague and with the building of the Romanesque St Vitus' basilica in the Hradčany. His successor Duke Vratislav I (1061–92) proved himself a true ally of the Emperor Henry IV, who, at the Synod of Mainz in 1085, appointed him King of Bohemia. The Archbishop of Trier officiated at his coronation in Prague in 1086.

Marketplace of Prague: The marketplace of

Left, group of figures in the Vyšehrad which portray the Libuše legend.

Prague is first mentioned in documents dating from the late 11th century. It was situated in the area of today's Staroměstské náměstí (Old Town Square). In the early 12th century this began to develop into a settlement in its own right, becoming known as the Old Town (Staré město). In the ensuing construction boom, the Romanesque style spread, as the many church foundations dating from this time show – St John's, the Holy Cross, St Laurence's, St Andrew's, St Leonard's, St Mary's, SS Philip and James' in the Bethlehem Square, St Clement's and St Aegidius', almost all of them in the Old Town.

Sobeslav I (1125–40) completed the alterations to the Vyšehrad, begun by Vratsilav II. The Romanesque churches and the castle citadel were completed and the clay walls replaced with stone fortifications. Sobeslav did all he could to encourage the expansion of the economy and of trade. Under his patronage, the Vyšehrad Cathedral Chapter produced the Vyšehrad Codex, a richly decorated manuscript.

Vladislav II (1140–72 and known as Vladislav I after his coronation as King of Bohemia) moved the ruler's residence to Hradčany. The new king had an aristocratic palace built, and the basilica of St Vitus was extended. The Premonstratensian monastery of Strahov was built in 1140, and in 1170 the St John's monastery was founded on the level of what was later to be the Malá Strana bridgehead of the Judith Bridge.

King Přemysl Otakar I (1198–1230) completed the building of the stone Judith Bridge (about the same level as today's Charles Bridge), begun in 1170, thus creating a permanent link between the castle and the Old Town, and providing a boon to further development. On the opposite bank, under the walls of Hradčany, he also founded the community known as Malá Strana (Lesser Quarter – literally "small side"). It became the home of skilled workers, carters and fishermen, and even today it remains a quarter inhabited mostly by ordinary people, students, writers and artists. Under Otakar's rule, Bohemia rose to become an important Central European power. Prague, residence of the Přemyslid dynasty, was promoted to one of the international meeting places of Central Europe. Long-distance trade had largely been re-routed along the Danube and through Vienna, but local needs and the luxury trade made up for the shortfall.

Extraordinary privilege: By 1230 the Old Town had been given borough status and, in 1231, King Wenceslas I (1230–53) set about defending it with a system of walls and fortifications. The building of a city wall was a sign of extraordinary privilege, as a wall meant more than protection against attacks from outside. The popular slogan of the times "city air makes for freedom" encouraged people from the outside to come and settle. If you lived in a town for a year, you automatically became one of the citizens.

The Old Town remained under the jurisdiction of a judge appointed by the king – his position was equivalent to that of a governor. It was administered by a council which met in the house of the judge. Not until around 1338 did John of Luxembourg grant the Old Town the right to its own town hall. The judge and the councillors were nominated from among the patrician families (nobility and land-owners) and elected by the members of those same families.

It is generally assumed that the first guilds and craft associations arose in the Old Town towards the end of the 13th century. The Jewish quarter was ruled directly by the king, and some church property maintained its special status.

The Malá Strana was granted privileges and rights equivalent to the Magdeburg Charter in 1257. Centred around the castle, Hradčany developed as the quarter of the aristocracy and clergy. It received a city charter and its own walls in 1320.

So at the beginning of the 14th century, Prague was composed of three different towns, the Old Town, the Malá Strana and Hradčany, each legally and administratively independent and also very different in their social and demographic structure. The total area of the city in 1300 is estimated at 297 acres (120 hectares), of which 198 acres accounted for the Old Town alone.

Incipit aurea bulla imperialium constitucionum. Et primo invocacio ad summum creatorem. Omnipotens eterne deus spes unica mundi. Qui celi fabricator ades qui conditor orbis. Tu populi memor esto tui si te mitis ab alto prospice ne gressum faciat ubi regnat erinnis. Imperat allecto leges dictante megera. Sed pocius virtute tua quem diligis huius Cesaris insignis Karoli deus alme ministra

Ut valeat ductore pio per amena virecta florencium semper nemorum sedesque beatas. Ad latices intrare pios ubi semina vite. Divinis animantur aquis a fonte superno. Fecunditata seges spinis mundatur a tempris. Ut messis queat esse dei mercisque future maxima centenum cumulare per horrea fructum

THE IMPERIAL CAPITAL

From the middle of the 13th century the fortunes of Bohemia and therefore of Prague had been determined by the political and military successes of King Přemysl Otakar II (1253–78). His power, which rested securely on the vast wealth derived from the silver mines of Bohemia, rose steadily until he reached a position of hegemony in Central Europe. He advanced into Hungary, into Slovakia and even pushed the frontiers of Bohemia as far as the Adriatic. He won the respect of both the Pope and the Emperor, but then died at the battle of Dürnkrut fighting Rudolf of Habsburg, who considered the Czech king his greatest rival. The subsequent crisis concerning the succession was partially alleviated when the claims of King Wenceslas II to the Polish throne were recognised. For a while Henry of Carinthia, Wenceslas II's brother-in-law, ruled, then it was the turn of Rudolf of Habsburg, then Henry again. The Přemyslid dynasty finally died together with the young Wenceslas III, who was assassinated in 1306 during his Polish campaign. He was the last of the line.

The times were marked by unrest and political power struggles, in which the patrician families of Prague took part, with varied success. The city was besieged, laid waste and plundered several times. In 1310, the Bohemian Estates chose John of Luxembourg (1296–1346) as their new king. On 31 August 1310 he had married Elizabeth, the younger daughter of Wenceslas II, and when he stood with his armies before the walls of Prague on 3 December that same year, the citizens offered no resistance. John was intensely – and expensively – involved in imperial politics and spent most of his time out of the country. The castle of Prague fell gradually into decay, but the Old Town was able to buy numerous privileges from John, among them the town hall mentioned in the previous chapter.

Preceding pages: extract from the "Golden Bull" of Charles IV. Left, Charles IV, from a votive tablet by Johann Ocko.

Emperor Charles IV: However, John's son, Charles IV, was particularly fond of Prague. Charles was born in 1316 and baptised Wenceslas, but at his confirmation he took Charlemagne as his personal patron and changed his name. In 1323, John brought young Wenceslas to Paris, where Petrus Rogerii – later Pope Clement VI – was entrusted with his education. In 1333, Charles, after a short period spent in Italy, came back to Prague as his father's governor, and had the dilapidated castle rebuilt according to French models. However, Charles did not immediately take up the regency (with the recognition of the Bohemian Estates) of the Bohemian crown lands until 1340. He was now acting for his father, who had gone blind. On the death of his father at the battle of Crécy in 1346 he was elected king of the Germans and was crowned king of Bohemia one year later. In 1355 he was crowned Holy Roman Emperor in Rome. He chose Prague as his residence and set about making it not only the political, but also the cultural hub of Central Europe.

Peter Parler – Imperial Architect: In 1344, Charles had already managed to use his good relationship with Clement VI to get Prague promoted from a bishopric to an archbishopric. In the same year the building of St Vitus' Cathedral began, over the remains of the former basilica. The cathedral was conceived as a triple-aisled nave church, in the French idiom. Charles obtained the services of the French architect, Matthias of Arras.

Following the latter's death in 1352, the masons' and sculptors' workshop of Peter Parler (1332–99) took over the building. Under his direction the famous triforium arcade was created, along with the choir, the South Tower, and particularly the vaulting, for which Parler was famous. The Wenceslas Chapel and the "Golden Door", both evidence of a synthesis of architecture and sculpture never before achieved, can be attributed wholly to Parler. After finishing work on St Vitus' Cathedral, Parler was contracted to complete the church of St Mary in the Týn.

Also designed and executed by Parler's workshop is the work around the windows of the Martinic chapel, the Bridge Tower of the Old Town and the church of the Charles Church, which was modelled on the chapel of Charlemagne in Aachen. In 1357, under Parler's direction, work began on the Charles Bridge. Peter Parler was one of the most notable architect of the Late Gothic period. His style, particularly his sculptural work and vault innovations, greatly influenced subsequent architecture.

The University: Prague gained considerably in cultural importance when the university – the first in Central Europe – was built. Charles IV granted the official founding Charter on 7 April 1348. The Charles University was intended to draw together scholars from all the regions of the empire and had a constitution similar to that of the University of Paris, where Charles had studied. It was divided into four "nations", Bohemian, Bavarian, Saxon and Polish. However, these did not represent actual national groupings, but symbolised the four points of the compass by giving them the name of the nearest national neighbour in that direction. The "nations" were important insofar as each had a vote in the University's decision-making process, and the posts of Rector and Chancellor were filled by each "nation" in turn.

To begin with, lectures were held in churches and in the "Lazarus House" in the Jewish Quarter. The move to the Carolinum, today one of the oldest university buildings in the world, did not take place until 1356. Charles, who was a writer himself and was one of the few educated medieval rulers, was able to gather leading thinkers and scholars of his time around him. Without doubt, Prague in the 14th century was one of the most important cultural centres of Europe. Among the early Humanist circle around Charles were people like Cola di Rienzi, Ernest of Pardubitz (later Archbishop of Prague) and the famous Johann of Neumark. He was Charles's chancellor and had considerable influence on the spread of the New High German language. He was famous as a translator of Latin prayers and Bible texts.

A masterpiece of civic planning: However, Charles made his most noticeable mark on the development of Prague by founding the New Town (Nové město) and thereby almost doubling the city's area. The New Town spread in a circle from the southeast of the

Old Town to the river below the Vyšehrad and to the Porici. This area incorporated some smaller settlements and several monasteries, among them the Carmelite convent with the church of St Mary of the Snows and the Emmaus monastery, which were founded in 1347 and already planned to fit in with the designs for the future New Town. Also in 1347, the foundation stone of the Charles Church was laid.

Neither the plans nor the name of the architect have survived, but an analysis of the basic town plan, which remained almost unaltered until well into the 19th century, shows clear evidence of creative and farsighted planning. The official founding charter was granted by Charles on 3 March 1348 – probably the day on which the foundation stone for the city walls was laid. Settlers who wanted to live in the New Town were assigned plots of land and, in return for tax concessions, had to complete their houses within 18 months.

The New Town was designed to fall into four large sections. The central section is constructed in a geometric plan around Wenceslas Square (the former horse market); the southern portion is centered on Charles Square (formerly the cattle market); and the north is dominated by the former merchant street of Hybernská. The lower area, the valleys and slopes of the former settlement of Slup, which were planted with orchards, fields and vineyards, was hardly built up at all.

Left, Charles IV with the imperial regalia. Above, the seal of the Charles University, the oldest university in Central Europe.

The generous scale of the plan is most remarkable. The Ječná (Barley Street) is nearly 89 feet (27 metres) wide, and Charles Square with its length of 1,706 feet (520 metres), an area of nearly 20 acres (8 hectares) is the largest square in Europe. However, Charles's building schemes were not confined to the extension of the castle and the New Town. On the left bank of the Vltava a new wall was built and the area of the Malá Strana increased considerably.

The Hradčany settlement was also fortified. Part of this fortification is known as the "Wall of Hunger", for the story goes that Charles ordered this wall to be built during a time of widespread poverty in order to reduce unemployment. At the same time, building began on the Petřín Hill. In the Old Town, too, the building boom that had begun in the 1330s continued. It provided the city with the rebuilding of the monastery and church of St James, the alterations to the churches of St Aegidius, St Martin, St Castullus, St Gall, St Nicholas and also the new building of the church of the Holy Ghost, to name only a few of the most important churches.

After the tremendous boom time of Charles's rule, Prague, at the end of the 14th century consisted of two castles and four "towns" with an area of 1,976 acres (800 hectares) and a population of over 50,000. Within the city area, there were around 100 monasteries, churches and chapels, several dozen markets and an impressive system of water supplies. Prague had been promoted to an imperial residence, an archdiocese, the seat of a papal legate, and a university town. The Czech money minted at the nearby silver-mining town of Kutná Hora served as the hard currency for the entire region. In uneasy alliance with the kings, foreign merchants, notably Germans and Italians, became economically and politically powerful. But social stability was undermined by the emergence of guilds of craftsmen which were often torn by internal conflicts. Increasing numbers of poor provided a further volatile element. The uneven social structure, caused in the first instance by the social and national contrasts among the settlers and also between the inhabitants of the Old and New Towns, resulted in a break between the two settlements in 1377. When Charles IV died on 29 November 1378, he left Prague with a lot of building sites, and also with a lot of problems smouldering under the surface.

Wenceslas IV: Charles's son, Wenceslas IV, succeeded his father's position in the Empire and in Bohemia. He faced strong political opposition within the Empire and had to accept a considerable loss of power and authority. Because of this, Prague decreased in importance and kudos. Building works soon slowed down, economic difficulties arose and led to a depression that brought social unrest in its wake. The dissatisfaction of large sections of the population, especially the poorer Czech inhabitants of the New Town, was focused on the rich (mostly foreign) patrician families and particularly on the clergy. In 14th-century Europe, a general opposition to the luxurious and often not very moral lifestyle of the monasteries was growing. In Prague this had already come under critical fire, occasionally from members of its own ranks, during the reign of Charles IV. In the late 14th century, Konrad Waldhäuser (died 1369) and Jan Milic Kromeriz (died 1374) were prominent preachers who attacked the luxury and immorality in the monasteries.

In 1391, the Bethlehem Chapel was founded. Its plain exterior alone (the chapel was rebuilt exactly to original plans in the years 1950–53) marks it as totally separate from the criticised conditions. In March 1402,

Jan Hus (1369/70–1415) began to preach in this chapel against the secularisation of the church.

The 45 Articles of the English theologian John Wycliffe (1328–84) had a considerable influence on the reforming work of Jan Hus, who urged a re-awakening of the church, based on a return to the message of the Bible and a lessening of the gap between clergy and laity. Hus's ideas were very popular among the citizens, and even with Wenceslas IV. However, the clergy, afraid of losing power, rejected him decisively.

Jan Hus: In 1398, Hus was appointed to the University as "magister primarius", as Professor of Philosophy. Here he continued to expound his ideas for reform. However, Hus was defeated in a vote by the "nations" – most of the academics condemned his and Wycliffe's thinking. In 1409, the dispute reached a crisis. It was no longer purely a theological quarrel, but now had nationalist overtones. This was the year in which Hus managed to obtain the Decree of Kuttenberg (Kutná Hora) from his patron Wenceslas IV. This granted the Bohemian "nation" in the University a majority of votes.

After making strong protests, the German academics moved out of Prague in a body and began an empire-wide campaign against the University of Prague. At the same time, the clergy of Prague reacted to Hus and his followers with arrests and repressive measures, as did the patrician families, who were greatly disturbed by the social and political applications of Hus's ideas as proposed by Jan Želivský.

Escalation of the conflict came on 30 July 1419, when an angry mob led by Želivský marched to the Town Hall in the New Town and demanded the release of the arrested Hussites. The consuls refused, and the enraged citizens stormed the Town Hall and began the tradition of defenestrations in Prague when they hurled the consuls and seven other citizens who were defending the Town Hall out of the window. Unrest spread rapidly, and could no longer be controlled even by royal troops. Soon the Hussites occupied the Town Hall and elected their own consuls. Wenceslas made no effort to supress them and, in August, approved the appointment of their consuls.

Perhaps he hoped that he could tone down the political conflict with this move, but the signal came too late. Wenceslas IV died on 16 August, 1419 in Nový Hrádec. On 17 August, the Hussites stormed the Carthusian monastery in Ujezd and continued the revolution.

Left, Jan Hus being led to the stake in Constance. Above, in the shadow of his father Charles: Wenceslas IV.

PRAGA

RUDOLF II AND THE HABSBURGS

The winners in the Hussite revolution were, for the most part, the Czech nobility. During the decade from 1430 to 1440 the Catholic city governors were driven out, church property was confiscated, Prague's independence was legally confirmed – triumphs for the Bohemian Estates. Only Czechs of Hussite persuasion were allowed to vote on the Prague city council. Rome agreed to religious freedom with the so-called *Four Articles of Prague*. However, the Catholic church was merely biding its time.

But for now the people of Prague acclaimed their first Hussite king. In 1458, George of Poděbrady was crowned. A Czech and an Utraquist (Utraquists were the more moderate Hussites), he was King of Bohemia for 13 years, and Prague, laid waste by the revolution, its economy in ruins, blossomed once more. George had the towers of the Týn church and the Bridge Tower in the Malá Strana built. He had no time to do more, as Rome incited people against the "heretic king".

During the rule of George's successor, the Polish prince Vladislav (1471), the second Defenestration of Prague occurred. Vladislav had let the Catholics back in. They occupied the Old Town Hall, arming themselves against the Utraquists. But then dark plots were made public. During the nights of 25 and 26 September 1483, the populace stormed the Town Hall and threw the spokesman and the mayor of the Old Town out of the window.

In 1484, Catholics and Utraquists made peace once more with the Treaty of Kuttenberg (Kutná Hora). The unrest had badly affected King Vladislav, whose residence was in the royal palace in the Old Town. He moved and chose the dilapidated castle as his new residence and renovated it. The masterpiece is the Vladislav Hall, the first hint of the Renaissance style among the Late Gothic prevalent in Prague, and was built by Benedikt Rieth (1454–1534).

Preceding pages: view of Prague *circa* 1493. Left, Van Dyck portrait of General Wallenstein. Above, Prague flourished once again under the rule of Rudolf II.

It was the teaching of Martin Luther that split the Utraquists into two camps during the first half of the 16th century. The Old Utraquists were against the German Reformation, the New Utraquists supported it. Catholicism also found support from Vladislav's son Louis, who began his reign in 1516. Following Louis's early death the divided Estates chose a new king in 1526. They made a fatal choice: the new King of

Bohemia was the Austrian Archduke Ferdinand I of Habsburg (1526–64). Prague became the most important prop of Viennese rule, and Rome found more and more support for its fight against the Utraquists. The Hussite era was facing its final defeat. Ferdinand soon quarrelled with representatives of the Bohemian Estates.

It was Ferdinand I's pleasure palace that finally helped the Renaissance to establish itself in Prague. The Belvedere, on a hill opposite the Hradčany, became the model for the palaces of the nobility. John of Lobkovic, for instance, had his palace built in the latest Italian style. Later, this building came into the hands of the Schwarzenberg family. It still bears their name today, and is now the Museum of Military History.

In 1555, Ferdinand had the hunting lodge Star Palace built on the White Mountain. By this time Prague had surrendered unconditionally to his rule. However, in 1546 open conflict broke out once more. Ferdinand went to war against the Protestant German princes. The Estates in Prague refused to support the King in a war against their religious brethren. However, the King returned victorious, with clearly one thought on his mind: revenge. On 1 July 1547, Ferdinand's mercenaries swooped down on Prague. Never again would the city deny him obedience.

Ferdinand took all the privileges won by the glorious Hussite revolution away from Prague. The city became a vassal of the Habsburgs. All public property had to be surrendered. And now the way was open for the return to power of the Catholic church in this Protestant country. In 1561, Ferdinand succeeded in appointing a new Bishop of Prague – the post had not been filled since the Hussite revolution.

However, Prague acquired new fame in 1583, although it was no longer politically independent, as the residence of Rudolf II of Habsburg. Many historians compare the Rudolfine era with the glorious reign of Charles IV. Life came back into the city: diplomats, political observers, adventurers, traders from all over the world, craftsmen, professional soldiers, musicians and many artists followed the ruler.

While Rudolf indulged his passion for collecting works of art, political and religious conflicts continued to seethe, hardly noticed by those in the castle. Meanwhile, Rudolf hoarded art treasures – paintings and drawings were his particular favourites. His favourite painters were Dürer and Peter Brueghel the Elder. If he couldn't get hold of the original, he had it copied by Jan Brueghel or Peter the Younger. Rudolf also amassed paintings by other leading artists of the Renaissance: Titian, Leonardo, Michelangelo, Raphael, Bosch and Corregio.

The Emperor was also interested in all kinds of curiosities and rare objects. An inventory records: "In the two upper compartments all kinds of strange sea fish, underneath them a bat, a box of four thunder stones (meteorites), two boxes of lodestones, and two iron nails, said to come from Noah's Ark, a stone that grows which was a gift from Herr von Rosenberg, two bullets taken from a Transylvanian mare, a box of mandrake roots, a crocodile in a bag, a monster with two heads..."

Despite his somewhat indiscriminate passion for collecting, Rudolf II undoubtedly made Prague into the "artistic treasure house of Europe". Unfortunately, much was later lost during the Thirty Years' War.

In 1578 the Jesuits began work on the great St Saviour's Church. The basic construction still adheres to the old Gothic pattern, but windows and sills, reliefs and the vaulting are built using a more modern style. The great church of the Jesuits was the symbol of growing Catholic strength in Prague. The conflict between Rudolf II and his brother, Archduke Matthias, was a long-awaited trump card for the Protestants. Rudolf, under pressure from Matthias, had to make concessions. In 1609, the Estates forced the king to issue the *Majestát* (Letter of Majesty) guaranteeing religious freedoms. In 1611, after a failed attack on Prague, Rudolf had to surrender the crown to his brother, and died barely a year later.

The Thirty Years' War: In all ages there have been places in the world where political observers can feel the pulse of the times beating faster than elsewhere, places where a zeitgeist is essentially created. Prague in the early 17th century must have been such a place. The gulf between the House of Habsburg and the Bohemian nobility, between Catholic and Protestant, was a reflection of the political situation throughout Europe. Indeed, the conflict here had a long tradition unparalleled anywhere else. No one should be surprised, then, that Prague was the place over which the storm clouds of war, which had been hovering threateningly over all of Europe, broke first.

"Follow the old Czech custom – throw them out of the window!" a voice from the crowd is supposed to have shouted on 23 May 1618. The representatives of the Bohemian Estates were enraged and stormed the Court Chancellery in the castle. Protestant churches were burning in the surrounding

Left, Prague's forces fight against the Swedes on the Charles Bridge (1648). Above, the murder of Wallenstein in Cheb (1634).

countryside. In 1617, the oppressor of the Protestants, Ferdinand of Styria, had been crowned King of Bohemia. Count Martinic, Governor Slavata and Secretary Philipp Fabricius fell nearly 55 feet (16.5 metres) into the castle moat, and all of them survived. Prague was in uproar. The Thirty Years' War began and Prague became the centre of the revolt.

A year after the Defenestration, the Bohemian Estates got rid of Ferdinand II and made Frederick V, Elector of the Palatinate, King of Bohemia. But Ferdinand II was a Habsburg Emperor based in Vienna, and he

was going to pay them back. On 8 November 1620, the combined armies of the Emperor and the Catholic League were drawn up on the White Mountain outside Prague, facing the army of the Bohemian Estates. The outcome of the battle was decided within a few hours. The army of the Estates fled behind the walls of Prague, and King Frederick to the Netherlands. Prague was defeated. For weeks enemy troops plundered the city and the damage ran into millions of gilders.

A dreadful revenge indeed. The presumed leaders of the revolt were arrested and executed. Even before the executions were carried out the Emperor had reinstated the Catholic clergy. On 21 June, 15 citizens of Prague, 10 members of the nobility and two citizens of other estates were executed in the Old Town Square. The heads of the 12 leaders were impaled the Bridge Tower in the Old Town, as a permanent warning – a cruel penance for the Defenestration. In the same year all non-Catholic clergy were forced to leave Prague. The Emperor had torn up Rudolf's Letter of Majesty with his own hands. In the years that followed many families emigrated.

During the following 30 war-torn years only a few new buildings were constructed in Prague. One of them was the palace of Albrecht von Wallenstein (or Waldstein) in the Malá Strana. Wallenstein, who had assisted in crushing the Bohemian revolt and had been appointed commander-in-chief of all imperial forces, had his Prague residence built between1624 and 1630 on a site that had formerly contained 23 houses, a brickworks and three gardens. The palace still stands, but Wallenstein's ambition of becoming the supreme authority in a united Germany was never realised: he was murdered in Cheb (Eger) in 1634.

It was from around this time that the baroque style began to emerge. Churches that belong to the baroque era are that of St Nicholas in the Malá Strana (1704–55), St Nicholas in the Old Town, the church of St Catherine in the New Town (1737–41) and the Holy Trinity Church in the New Town (1720). The most famous baroque architect in Prague was the Bavarian Kilian Ignaz Dientzenhofer, who died in 1754.

Under the rule of Maria Theresa (1740–80) and Joseph II (1780–90) religious freedom returned to Prague. Joseph II proclaimed the *Edict of Tolerance* in 1781. The age of religious wars was at an end, but freedom of religion was not the only aim of the Bohemian Estates.

Above, Joseph II, the son of Maria Theresa. Right, Maria Theresa, Empress of Austria and Queen of Bohemia (1717–80).

THE ROUTE TO THE REPUBLIC

After the Thirty Years' War, any last glimmer of post-Hussite Czech national consciousness was extinguished at the battle of the White Mountain in 1620. Twelve heads hung on the Bridge Tower, gruesome symbols of the destruction of Czech culture.

In the following centuries, the Czech language disappeared from the Estates, and the Czechs became a people of peasants, small-time craftsmen and servants. In the 18th century, the upper classes, nobility and bourgeoisie, were German. Language differences often made communication between people and administrators impossible. Maria Theresa commanded that all justices and civil servants should know the "vulgar tongue". Teachers taught Czech once again.

The Czech Language Comeback: The late 18th century was the great period of the theatre in Prague. The first Czech performance was held as early as 1771. In 1781, the Nostitz Theatre was built by Count Anton Nostitz-Rieneck (today it is the Estates Theatre). When the theatre opened, the German upper classes went to see Lessing's *Emilia Galotti* or, in 1787, applauded Mozart's *Don Giovanni*. While this was happening, the Czechs struggled to put on matinées. In 1785, they gained permission to do so for a short time, but soon Czech performances in the Nostitz Theatre were banned once more.

The Czech players moved to the Bouda (booth) in the Horse Market. This was a little wooden theatre. It was to be another 60 years before the Czech nation was to appear on the stage of history, but the Czech language could no longer be suppressed. Czech books and newspapers returned, and Czech was again taught at the University.

In 1833, the Englishman Edward Thomas began the production of steam engines in Karlín. The rapid economic development of the Industrial Revolution created an industrial proletariat in and around Prague and the number of Czechs in the population of Prague grew. Tension increased between Germans and Czechs. However, at first both had a common enemy: the all-powerful Viennese State Chancellor Metternich. The year of revolution, 1848, was the last time that Germans and Czechs together manned the barricades for a common cause.

Even then the goals were not really common to both parties, for the Czechs were no longer really interested in Bohemian, but in Czech freedom. In February 1848, there was revolution in Paris, and Metternich resigned in Vienna on 13 March. There was rejoicing in Prague. But the rejoicing was divided – the Germans wanted to accept the invitation of the revolutionary Frankfurt Parliament, the Czechs wanted their own state as part of a federal Austrian Empire.

Violence at Slavic Congress: The Slavic Congress met on 2 June 1848 in the museum building of Prague. One of its demands was for equal rights for all nationalities. The leader of the movement was the Czech František Palacký: "Either we achieve a situation where we can say with pride: 'I am a Slav', or we shall stop being Slavs". The Congress came to a violent end. After a Slavic mass, the Prague militia fired into the crowd, and the nationalist movement was crushed in bloody fighting in the streets and on the barricades.

From 1849 on, the citizens' revolutions in all the nations of the Austrian Empire came under the rod of Absolutism. The Czech language was even forced back out of the civil service departments. But after the war of France and Italy against the Austrian monarchy, Absolutism was at an end. On 5 March 1860 an "extended imperial council" was called. Representatives from different lands sat with the councillors appointed by the Emperor. In 1861, a Czech could become mayor of Prague, and yet on a national level the Emperor favoured the German-speaking Bohemians. The Czechs remained loyal to the Emperor in the struggle with Prussia in 1866 for supremacy in Germany, and yet

Right, Tomáš Masaryk, a leading Slavic nationalist, chief founder and first president of the Czechoslovak Republic (1918).

they were not rewarded. There was no Czech equality, never mind autonomy.

"Libuše" in the National Theatre: The Czechs could only win minor battles – for instance, they finally got their long-awaited Czech national theatre in Prague. It was opened on 15 June 1881 with a special performance of Smetana's *Libuše*. Unfortunately the theatre burned down on 12 August of the same year, but it was quickly rebuilt and remains a symbol of Czech national sentiment even today. In 1876, work began on the Rudolfinum, and from 1893 on the Bohemian National Museum was to be found in the Horse Market, today Wenceslas Square. The university was split in 1882 – there was now a Czech and a German university.

The Republic is born: But with the arrival of World War I, the Czechs were still fighting the Germans for equal rights. The war itself increased the estrangement, as the Germans fully supported the war effort of the Central Powers and the Czechs opposed it. Tomáš G. Masaryk, who had risen to prominence representing Czech national causes in Vienna, now went into exile to lead the campaign for independence. Together with his former student Edvard Beneš and the Slovak astronomer Milan Štefánik, he devised a programme of political union between the Czechs and the Slovaks. He established contacts with his countrymen living in Allied and neutral countries, especially the United States, and soon the idea began to gain ground. When Czech exiles then began fighting for the Allies, President Wilson gave his backing to "autonomous development" for the peoples of Austria-Hungary. A declaration favouring political union of the Czechs and Slovaks was issued at Pittsburgh on 31 May 1918.

After the recognition of the Czechoslovak National Council by France, other Allies soon followed and on 18 October a simultaneous declaration of independence was issued by Masaryk and Beneš in Washington and Paris. The Habsburg monarchy was on the point of collapse and had no choice but to accept the terms. The Czechoslovak Republic was proclaimed by the Prague National Committee on 28 October, a move that was repeated by the Slovak National Council two days later.

Above, shop in the Jewish quarter, before 1900. Right, Edvard Beneš, state president of the First Republic and again after 1945.

RÖSSLER
Droguist

L. HAUROWITZ

On 28 October 1918, the Czechoslovak Republic was declared in Prague. But the borders were still not confirmed: the Sudeten Germans wanted to join with German-speaking Austria, the inclusion of Slovakia in the Czech republic had not yet been decided, and Poland was claiming the coal mines in the former duchy of Teschen.

On 14 November, Tomáš Masaryk was elected President of the Republic and was welcomed back by the enthusiastic people of Prague after four long years of exile. Foreign Minister Edvard Beneš – Masaryk's successor in presidential office – made skilful use of the last weeks of the war and the time after Germany's surrender. The uneasy powers of the Entente had no clear idea of how Europe should look once peace was declared. In 1918, the Czech foreign minister managed to obtain the incorporation of Slovakia into the Czech state, against Hungarian opposition and the opposition of the populace, who also wanted autonomy. Continual internal unrest was guaranteed by demands for "national autonomy for Slovakia". The incorporation of the Sudetenland had even worse consequences. Later, it gave Adolf Hitler a welcome excuse to liquidate Czechoslovakia. Following the formula of the president of the United States, Woodrow Wilson, regarding the right of peoples to self-determination, representatives of the Sudeten Germans had declared an "autonomous province of the state of German Austria" on 28 October 1918. Towards the end of that year, Czech troops retaliated by occupying German-settled areas. The Peace Conference decided in favour of Czechoslovakia. Without a plebiscite, the German areas went to Czechoslovakia. Those parts of Teschen which were incorporated were not all in favour, either. After the Treaty of Versailles, opposition arose all over the country.

The Sudeten Germans saw themselves as an oppressed minority, disadvantaged by language rulings, land reform, and the unfavourable position of the German education system and industry. The economic effects of "Black Friday" on 4 October 1929 strengthened radical opinion. Around two-thirds of the 920,000 unemployed in the winter of 1932–33 were Germans. In 1933, the gymnastics teacher Konrad Henlein founded the "Sudetendeutsche Heimatfront" (SHF = Sudeten German Home Party). In 1935 the SHF, now the "Sudetendeutsche Partei" (SdP = Sudeten German Party) took part in the elections and soon became the voice of the people in the German-settled areas. It did not take long for Henlein to make contact with Hitler. After a few years he and his party became puppets of their mighty patron. This was why the SdP's demands became more and more radical over the years, and why their goals shifted from equal rights, more and more openly, to inclusion in the German Reich. In 1937, Henlein drew the conclusion "that today even the broad mass of Sudeten Germans no longer believe in equal rights with the Czech people in a Czech state."

In the autumn of 1938 the time was ripe. Afraid of war, France and Britain gave in to pressure from Hitler. On 30 September 1938, Chamberlain, Daladier, Mussolini and Hitler signed the Munich Agreement. The Sudeten lands now belonged to Germany. On 22 October, Beneš went into exile in England, and on 14 and 15 March 1939, the fate of Czechoslovakia was sealed. Hitler declared a "sovereign" Slovakian vassal state and established the Protectorate of Bohemia and Moravia. German troops marched into Prague on 15 March without meeting any form of resistance. The time of worst oppression began. The chief of the SD (Sicherheitsdienst = security forces), Reinhard Heydrich, attacked the Czech intelligentsia and the middle classes. After Heydrich's assassination by the Czech resistance in 1942, this process was continued by General Kurt Daluege.

The Czechs were defined as second-class persons under Article 2 of the Treaty of the Protectorate. Czech universities were closed. Academics were not allowed to work in their professions. Thousands were arrested and imprisoned in the concentration camps of Dachau and Oranienburg.

Preceding pages: Wenceslas Square, before 1900. Left, Sudeten German women acclaim Adolf Hitler. Above, crossing the border to Czechoslovakia.

FROM PEOPLE'S DEMOCRACY TO SOCIALIST REPUBLIC

In post-1945 Czechoslovakia, the path to socialism had been smoothed by history as in no other country. Before World War II, the country was one of Europe's highly developed nations, with modern light and heavy industry, and efficient agriculture. Above all, it possessed a self-confident, highly qualified proletariat and a class of educated people, intellectuals and artists, who were, if not actually members of the Communist Party of the time, nonetheless mainly left or liberal-left in their views. The relationship of the Czechs to the Soviet Union and to socialism was positive. Following the rebirth of Czech national consciousness in the early 19th century, the Czechs viewed the Russians as a kind of Slavic older brother.

Apart from two exiled groups, one in England and another group in Russia, the Czechs did not fight against Nazi Germany in World War II. The number of victims that they mourned in 1945 was small compared with those from Poland, Russia or the Ukraine. However, no other people in Europe had lost such a high percentage of their intellectuals and artists under Nazi rule. In the years from 1939 to 1945, the Nazis in the Protectorate of Bohemia and Moravia systematically exterminated not the common people, not the workers, not even the technologically qualified, but the academic and creative elite of the nation.

When World War II ended in May 1945, the Czechs enthusiastically greeted the soldiers of the Red Army (who had occupied much of Czechoslovakia) as liberators and Slavic brothers.

In February 1948, the communists gained control of Czechoslovakia in a bloodless coup. At that time they could still count on the support of the majority of Czechoslovak people. However, bitter disillusionment was just around the corner.

Left, announcement to the crowds in the Old Town Square. Above, the sudden end of the "Prague Spring" – the dream of a humane form of socialism is over.

The Stalin Years: The name of the disillusionment was Josef Stalin, who mercilessly forced the Czechoslovaks to accept his version of communism. From 1948 onwards, the Czechoslovak Communist Party, by now Stalinist in orientation, succeeded in destroying any idea of a specifically Czechoslovak way to socialism. In 1960, the country's official title of "people's republic" was changed to "socialist republic".

The rule of the Stalinists, first under Gottwald and then under Novotný, had dire

consequences for Czechs and Slovaks. In the late 1950s and early '60s scepticism and cynicism spread throughout Czechoslovakia. Hardly anyone still believed in Marxism or Leninism, or even in the idea of just and fair socialism. Also, Czechoslovakia was at rock bottom, economically speaking, by 1963. The events, five years later, which came to be called the "Prague Spring", had already begun in 1963 with the failure of the so-called Five Year Plan and the near collapse of the whole economy of the country.

Literature and Politics: In those years before the Prague Spring of 1968, a literature independent of the previously all-powerful Party censorship machine was developed in Czechoslovakia. Literature is of particular importance because, ever since the rebirth of the Czech national consciousness in the early 19th century, Czech writers have had a vital role to play.

When the crisis of Czechoslovak society and socialist beliefs came, in the early 1960s, it was once again authors such as the future Nobel Prize winner for Literature, Jaroslav Seifert, the lyric poets Vladimír Holan, František Halas and others, who replaced the helpless functionaries from Party headquarters as political and moral institutions. By 1965, at the latest, the Czechs and Slovaks had realised that this form of socialism could not continue, and, searching for something to cling to, they rediscovered their poets and authors. "Socialism with a human face" was a programme that was developed not by the Communist Party and not by the exhausted and insecure ideologies, but by the poets of the 1968 Prague Spring.

End of a Dream: And "Socialism with a human face" was indeed what Alexander Dubček claimed to be offering. The Action Programme embodied a number of reform ideas which included federal autonomy for the Slovaks, long overdue industrial and agricultural reforms, a revised constitution that would guarantee civil rights and liberties and democratisation of the country and the party. Behind these proposals lay a bitter admission and an undertone of despair: up until the spring of 1968 the face of socialism in Czechoslovakia had obviously not looked very human at all.

The dream of an efficient, just and happy socialist society came to a final end in Czechoslovakia with the collapse of the Prague Spring in 1968. In that year, the Soviet Union marched in with soldiers from five socialist countries and destroyed the Czechoslovak hope of national independence. They left behind them the power structure of a totalitarian, Stalinist state, which, although it called itself Socialist, had little to do with the ideal of true socialism, which the Prague reformers under Alexander Dubček had wished to realise.

With the purges of the Husák regime, many writers, composers, journalists and historians, as well as scientists found themselves unemployed and forced to accept menial jobs to earn a living. Those who tried to continue the struggle were silenced, but despite the indifference of the mass of the population, the discontent continued.

It erupted again in January 1977, when a group of intellectuals signed a petition, known as Charter 77, in which they aired their grievances against the Husák regime. The spokesman of the group, Václav Havel, had already written, in an open letter to President Husák two years before: "You have chosen the path that is the most convenient for you and the most dangerous for the country: the path of maintaining external appearances at the cost of internal collapse… the path of merely defending your power, at the cost of deepening the spiritual and moral crisis of this society and the systematic erosion of human values."

Above, State president Husák. Right, Alexander Dubček in 1968. His vision of "socialism with a human face" was short-lived.

Účastníci 1. světové války v průvodu

"Havel na Hrad" (Havel for the Castle) – loudhailers, posters and banners clearly expressed the wishes of the people of Prague. After the massive changes in other Eastern Bloc countries in the late 1980s, the power-hungry Czech government couldn't save itself either. The citizens of Prague demonstrated in Wenceslas Square in their hundreds of thousands and, despite the obstinate stance of the authorities and the large contingents of police and security forces, their voice could not now be silenced. Remarkable though it would have seemed just months before, one of the last major communist governments was about to topple.

Václav Havel, a co-signatory of Charter 77, was a central figure in these protests. The playwright, poet and political disident had already been a prominent participant in the Prague Spring of 1968. Following the Soviet clampdown that same year, his works were banned and his passport was confiscated. During the 1970s and '80s he was repeatedly arrested and served four years in prison for his activities on behalf of human rights.

A forum for change: The Civic Forum, a coalition of old and new opposition parties, was founded in November 1989. Havel, along with Jiří Hajek, who had been Foreign Minister under Dubček, was made a spokesman of the movement. Old and new critics of the regime now stood together in a united front. On 29 December 1989, Havel was unanimously elected President by the parliament in Prague. Alexander Dubček, veteran leader of the Prague Spring, was voted president of Parliament. After the peaceful revolution in East Germany, a "velvet revolution" now took place in Prague.

In the summer of 1990, in the first free elections for two generations, the Civic Forum received the endorsement of most of the electorate. But like most revolutions, even velvet ones, its leaders gradually drifted apart as the common enemy disappeared. The differences of opinion surfaced during debates on lustration, the "outing" of those who had held any high office in the Communist Party or who had any connection with the secret police. It was meant not only to be a spiritual cleansing, an atonement by a handful for the rest of society, but also a practical measure to bar tainted officials from public office for five years.

The eclipse of Václav Havel: Tthe key question was how to achieve the economic reforms that would lead to a free market. Throughout 1990 and early 1991, the great hope was foreign investment. When this failed to materialise on the scale required, an

ambitious coupon scheme was introduced, giving every adult the chance to buy shares in the state firms being privatised.

The enormous success of the scheme boosted support for Václav Klaus, the country's Finance Minister and champion of the free market. After the election of June1992, this sharp-witted speaker became the most important political figure in Czechoslovakia, eclipsing even Havel. It was Klaus who ended up having to deal with the split between the Czechs and the Slovaks, a situation which Havel had spent two years in office trying desperately to prevent.

Going separate ways: When the economy began its shift away from socialist planning after the revolution, the Slovaks were on the receiving end. Economically irrational but emotionally predictable, Slovak separatism is a symptom of uncertainty, a force liberated by changed circumstances at a time when the state's institutions are not strong enough to contain it. The 1992 elections produced a clearer result than expected. Around 34 percent voted for Klaus's ODS in the Czech Republic and Meciar's HZDS in Slovakia. The people had chosen two strong leaders with two different higher ideals, fast economic reform versus a sovereign Slovakia able to dictate a slower pace.

It quickly became clear that the only thing they could agree on was to split the country, even though neither party had been elected on this mandate and a majority of people, both Czechs and Slovaks, were actively against it. The Slovak Prime Minister, Vladimir Meciar, is a Moscow-trained lawyer. The growth of his power is seen by many in Prague not as proof that democracy works but a threat to its very existence.

President Havel's fate was also sealed by the election. Days before it, he gave a veiled warning against Meciar's style of political demagogery. When elected, Meciar did not forgive him and his party did not vote him back into office. When the Slovak parliament passed its own constitution in August, Havel resigned. However, that is by no means the end of this poet turned politician: he has indicated that he might accept the presidency of the new independent Czech Republic set to be formed in January 1993.

Václav Klaus leads the only genuine right-wing government in post-communist Europe. The average Czech, he could point out, is more pro-European than a Dane, has more respect for his government than the average Italian, and even (like the British) voted for right-wing reform in the teeth of recession. The Czechs, he says, are marching firmly on the road back to Europe.

Preceding pages: a dialogue with history is again taking place. Left, Václav Havel's dreams of unity were not fulfilled. Above, housework has begun in Hradčany.

Piwowar
u
Fleku

HP
MADE I

Neither the Czechs nor the Slovaks had considered living together in one state until the end of World War I. The Czechoslovak National Council was formed in 1916 by Czech and Slovak exiles in the US, but not until the Habsburg monarchy was unquestionably collapsing did the Slovaks in America accept the ideas of the leader of the Czechs in exile at that time, Professor T. G. Masaryk (later president of Czechoslovakia), and sign the "Pittsburg Convention" on 31 May 1918. In the future republic of Czechoslovakia, Czechs and Slovaks were to have equal individual rights.

Up until 1918 both Czechs and Slovaks lived under the Habsburg monarchy, though Slovakia was counted as part of Hungary. The "Hungarianisation" of Slovakia was carried out with such thoroughness and brutality that by the beginning of the 18th century the Slovak language had ceased to exist. Under the influence of the rebirth of Czech nationalism the Slovaks re-discovered their "lost" language and developed it further. After 1918, the Czechs and the Prague government tried hard to build up the Slovak education system and to help the Slovaks build up their own intelligentsia. There were Czechs working in Slovak schools and colleges, in administration, in the judicial system and in other departments.

Of course, this led to misunderstandings – the Slovaks felt patronised and later even oppressed by the Czechs. However, the Czechs also contributed to this misunderstanding. The "Pittsburg Convention" of 1918 stated that the Slovak people were an equal partner in the new republic, but this was soon forgotten in Prague, and many Czechs looked on Slovakia as their colony.

Discontent in Slovakia found its expression in the programme of the Slovak separatists, the Slovak Populist Party. In March 1939, when the rest of Czechoslovakia was occupied by Hitler, the Slovak separatists felt that their moment in history had come. Under Monsignor Tiso, they split off from the republic and, under Hitler's "protection", formed an "independent" fascist Slovak state, allied to Nazi Germany.

By the end of August 1944, the Soviet army had reached the northern borders of the Slovak state. Slovak patriots, together with the communists, organised a popular revolution, which was crushed by the Germans but was of historic importance to the Slovaks. The Czechs and Moravians did not rebel against the Nazis until Germany had already surrendered.

The differences continue today. After the parliamentary elections of 1992, nationalistic tendencies have once again got the whip hand. It is certain that the Czechoslovakia of the future will not match the one envisioned by Masaryk and Havel. There only remains the hope that the future constellation of two separate republics will reflect the wishes and aspirations of the nationalities concerned.

Preceding pages: inside the venerable "U Fleků"; Hradčany at night; young hopeful for the ice hockey team. Above, Czechs and Slovaks demonstrate for a cleaner environment. Right, group in traditional dress in front of an inn.

Vchod do sálu

DOMI
RESURREC
Statio ad S. Mariam majorem.
Introitus.
Resurrexi; & adhuc tecum sum, Alleluja: posuisti super me manum tuam, Alleluja: mirábilis facta est sciéntia tua, Allelúja, allelúja.
Dómine probásti me, & cognovísti me: tu cognovísti sessiónem meam, & resurrectiónem meam.
Glória Patri, & Filio, & Spirítui sancto. Sicut erat in princípio, & nunc, & semper, & in sæcula sæculórum, Amen.
Oratio.
Deus, qui hodiérna die per Unigénitum tuum, æternitátis nobis áditum devícta morte reserásti: vota nostra, quæ præveniéndo aspiras, étiam adjuvándo proséquere. Per eumdem Dóminum nostrum Jesum Christum Fílium tuum: Qui tecum vivit & regnat in unitáte Spíritus sancti Deus, per ómnia sæcula sæculórum. Amen.
Léctio Epístolæ beáti Pauli Apóstoli ad Corinthios.
Fratres, Expurgáte vetus ferméntum, ut sitis

301
CA
ONIS
epulémur : non in
, neque in fermén-
equitiæ : ſed in ázy-
& veritâtis.
æc dies, quam fe-
exultémus, & læté-
Confitémini Dó-
bonus : quóniam
ricórdia ejus. Alle-
Paſcha noſtrum
Chriſtus.
ctimæ Paſchâli lau-
Chriſtiâni. Agnus
Chriſtus innocens
t peccatóres. Mors
nflixére mirándo :
us regnat vivus.
, quid vidiſti in
n Chriſti vivéntis :
eſurgéntis. An-
udárium & veſtes.
is ſpes mea : præ-
alilæam. Scimus
iſſe à mórtuis ve-
tor Rex miserére.
nur uſque ad Sab-
cluſivè.
ncti Evangélii
Marcum.
: María Magda-
a Jacóbi, & Saló-

Slavia
17–23
zav. den

THE FIVE TOWNS OF PRAGUE

Today you can no longer talk about the Five Towns, for Prague has now been divided into 10 districts. However, most people are interested only in the five historic towns: Hradčany and the Staré město (Old Town), the Malá Strana (Lesser Quarter) and the Nové město (New Town), as well as the Prague Ghetto, somewhat euphemistically known as Josefov ("Josephstown").

At the beginning of the 19th century, 80,000 people lived in the city. Gradually more districts were added, and the population grew steadily. Vyšehrad, Holešovice and Bubenec brought the population of the city to around 200,000 by 1900. By 1922, the area of Greater Prague had a population of 676,000. After World War I, the city grew by leaps and bounds. Its area tripled to a size of 193 sq. miles (550 sq. km).

New suburbs such as Severní město (North Town) with a population of 80,000 and Jižní město (South Town) with a population of 100,000 were built. The south-western suburb of Jihozápadní město, intended for 130,000 inhabitants, is still being built. Today about 1.3 million people live in Prague, and about 10 percent of Czech industry is based there.

City administration was not unified until the rule of Emperor Joseph II. The separate town halls are reminders of previous autonomy. Even today, each individual district of the city has its own special appeal. Hradčany castle and its surroundings, such as the Loreto church or the Strahov monastery, have a distinct atmosphere. The Malá Strana and the island of Kampa with its ostentatious palaces, built in the shadow of the rulers of Hradčany castle, is another, quite separate tour. Its wine bars, the lovely gardens of the Petřín hill, the view of the Vltava and the Charles Bridge by night – they all have a romantic fascination.

In earlier times, the inhabitants of the congested and dirty Old Town and the Jewish quarter must have felt envious when they looked across to the other bank of the Vltava. If you take a look from the beautifully restored Old Town Square at the narrow alleys and courtyards surrounding it, you will get a brief glimpse of old Prague, the Prague of Franz Kafka. Today the Pařížská is a splendid street in which companies such as Dior and Lufthansa have established themselves. But just go a few blocks further on.

The generous, well-planned and far-sighted designs of Charles IV and his architects are apparent when you walk through the New Town. This was planning well ahead of its time. The broad open spaces such as Charles Square or Wenceslas Square are products of a time when no-one had yet dreamed of cars or trams.

Preceding pages: shop window displays; Café Slavia; Charles Bridge with swans. Left, Letná Hill offers a remarkable view of the Vltava and the bridges of Prague.

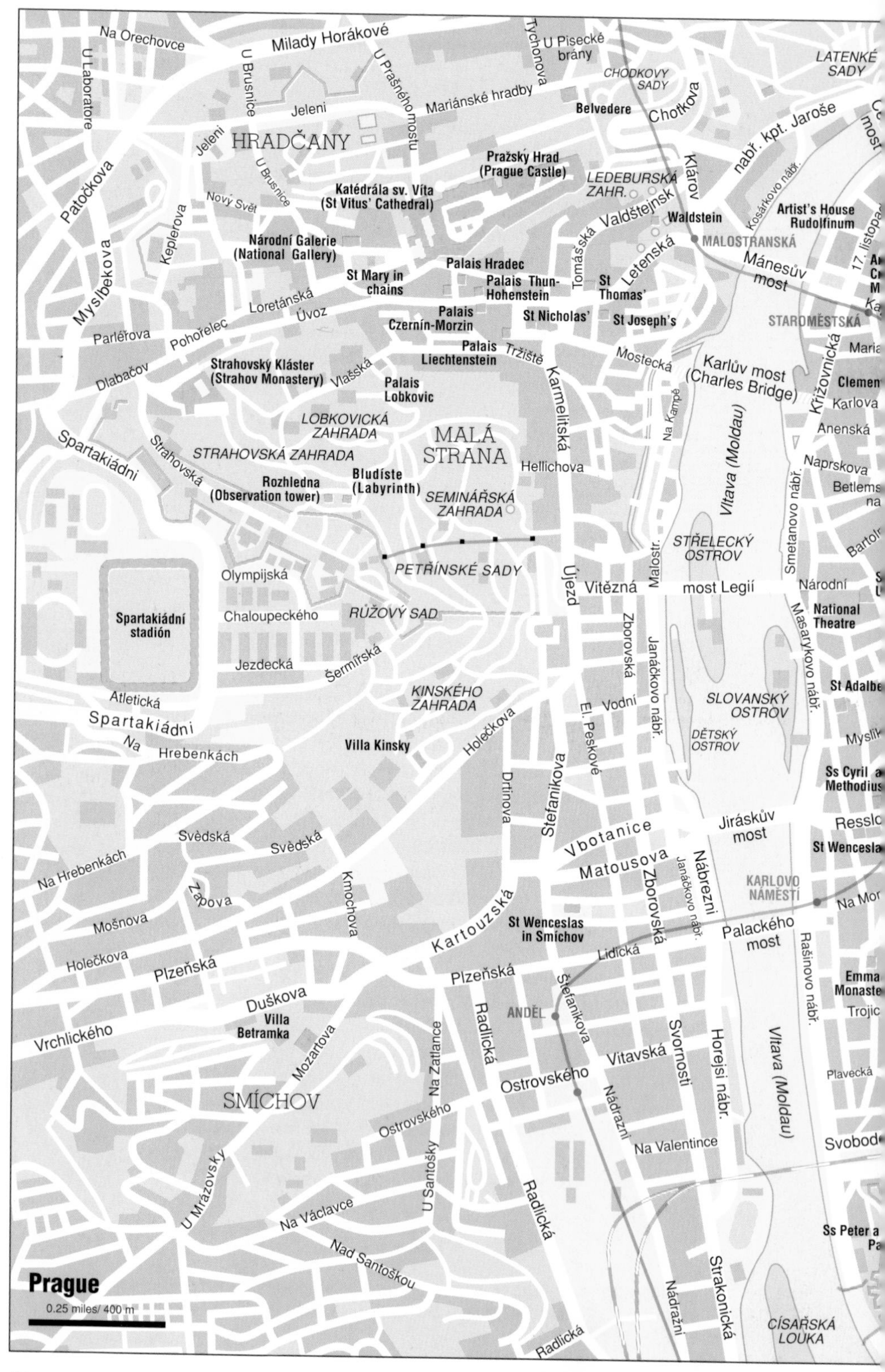

Prague
0.25 miles/ 400 m
HRADČANY
MALÁ STRANA
SMÍCHOV
Pražský Hrad (Prague Castle)
Katédrála sv. Víta (St Vitus' Cathedral)
Národní Galerie (National Gallery)
St Mary in chains
Palais Hradec
Palais Thun-Hohenstein
Palais Czernín-Morzin
Palais Liechtenstein
Palais Lobkovic
St Nicholas'
St Thomas'
St Joseph's
Belvedere
Waldstein
Strahovský Klášter (Strahov Monastery)
Rozhledna (Observation tower)
Bludíste (Labyrinth)
Spartakiádní stadión
Villa Kinsky
Villa Betramka
St Wenceslas in Smíchov
Artist's House Rudolfinum
National Theatre
St Adalbe
Ss Cyril a Methodius
St Wencesla
Emma Monaste
Ss Peter a Pa
LOBKOVICKÁ ZAHRADA
STRAHOVSKÁ ZAHRADA
SEMINÁŘSKÁ ZAHRADA
PETŘÍNSKÉ SADY
RŮŽOVÝ SAD
KINSKÉHO ZAHRADA
LEDEBURSKÁ ZAHR.
CHODKOVY SADY
LATENKÉ SADY
STŘELECKÝ OSTROV
SLOVANSKÝ OSTROV
DĚTSKÝ OSTROV
CÍSAŘSKÁ LOUKA
Karlův most (Charles Bridge)
Mánesův most
most Legií
Jiráskův most
Palackého most
Vltava (Moldau)
MALOSTRANSKÁ
STAROMĚSTSKÁ
KARLOVO NÁMĚSTÍ
ANDĚL
Milady Horákové
Na Orechovce
U Laboratore
U Brusnice
U Prašného mostu
Tychonova
U Písecké brány
Mariánské hradby
Jeleni
Chotkova
nábř. kpt. Jaroše
Klárov
Valdštejnská
Tomášská
Letenská
Kosárkovo nábř.
17. listopadu
Patočkova
Myslbekova
Keplerova
Nový Svět
Loretánská
Úvoz
Pohořelec
Parléřova
Dlabačov
Vlašská
Tržiště
Mostecká
Karmelitská
Na Kampě
Křižovnická
Karlova
Anenská
Naprskova
Betlems na
Smetanovo nábř.
Bartol
Národní
Spartakiádní
Strahovská
Hellichova
Malostr.
Olympijská
Chaloupeckého
Jezdecká
Šermířská
Újezd
Vitězná
Zborovská
Janáčkovo nábř.
Vodní
Masarykovo nábř.
Atletická
Na Hrebenkách
Holečkova
El. Peškové
Drtinova
Stefanikova
Vbotanice
Matousova
Nábrezni
Myslík
Resslo
Na Mor
Svèdská
Zapova
Mošnova
Kmochova
Kartouzská
Lidická
Plzeňská
Duškova
Vrchlického
Mozartova
Na Zatlance
Radlická
Štefanikova
Vitavská
Svornosti
Horejsi nábr.
Rašinovo nábř.
Trojic
Plavecká
Ostrovského
Nádrazni
Na Valentince
Svobod
U Mrázovsky
U Santošky
Na Václavce
Nad Santoškou
Strakonická
Nádražni

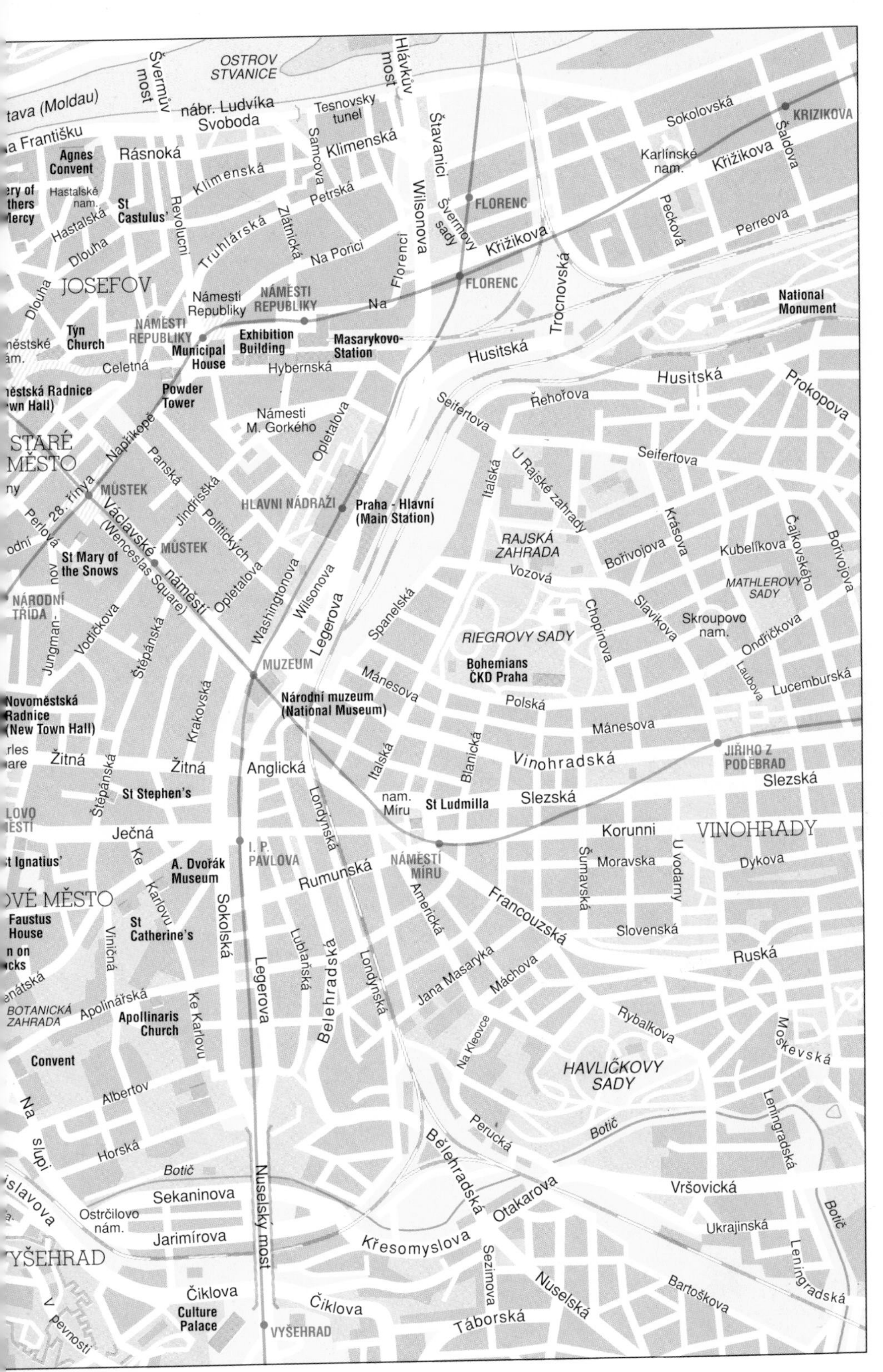

OSTROV STVANICE
Švermův most
Hlávkův most
tava (Moldau)
nábr. Ludvíka Svoboda
Tesnovsky tunel
a Františku
Agnes Convent
Rásnoká
Samcova
Klimenská
Štavanici
Sokolovská
KRIZIKOVA
Karlínské nam.
Křižikova
Šaldova
Hastalské nam.
St Castulus'
Revoluční
Klimenská
Petrská
Wilsonova
Švermovy sady
FLORENC
Pecková
Perreova
Hastalská
Dlouhá
Truhlářská
Zlátnická
Na Poříčí
Florenci
Křižikova
Dlouhá
JOSEFOV
Námesti Republiky
NÁMĚSTI REPUBLIKY
Na
FLORENC
Trocnovská
National Monument
Týn Church
NÁMĚSTI REPUBLIKY
Municipal House
Exhibition Building
Masarykovo Station
Celetná
Hybernská
Husitská
Husitská
Prokopova
Powder Tower
Seifertova
Řehořova
Námesti M. Gorkého
Opletalova
STARÉ MĚSTO
Na Příkopě
Panská
Seifertova
Italská
U Rajské zahrady
MŮSTEK
Jindřišská
HLAVNI NÁDRAŽI
Praha - Hlavní (Main Station)
28. října
Politických
Perlova
Václavské náměstí (Wenceslas Square)
MŮSTEK
RAJSKÁ ZAHRADA
Krásova
Čajkovského
Bořivojova
Kubelíkova
Bořivojova
St Mary of the Snows
Opletalova
Washingtonova
Wilsonova
Vozová
MATHLEROVY SADY
NÁRODNÍ TŘÍDA
Legerova
Spanelská
Chopinova
Slavíkova
Skroupovo nam.
Ondříčkova
Vodičkova
Jungmannova
Štěpánská
RIEGROVY SADY
MUZEUM
Bohemians ČKD Praha
Laubova
Lucemburská
Mánesova
Krakovská
Národní muzeum (National Museum)
Polská
Novoměstská Radnice (New Town Hall)
Mánesova
Žitná
Žitná
Vinohradská
JIŘIHO Z PODĚBRAD
Štěpánská
Anglická
Italská
Blanická
Slezská
St Stephen's
Londýnská
nam. Míru
St Ludmilla
Slezská
Ječná
Korunni
VINOHRADY
Ke Karlovu
I. P. PAVLOVA
NÁMĚSTÍ MÍRU
Šumavská
Moravska
U vodarny
Dykova
St Ignatius'
A. Dvořák Museum
Rumunská
Americká
Francouzská
NOVÉ MĚSTO
Sokolská
Faustus House
Viničná
St Catherine's
Slovenská
Lublaňská
Londýnská
Jana Masaryka
Ruská
Legerova
Máchova
Bělehradská
BOTANICKÁ ZAHRADA
Apolinářská
Apollinaris Church
Ke Karlovu
Rybalkova
Na Kleovce
Moskevská
HAVLIČKOVY SADY
Convent
Albertov
Leningradská
Na slupi
Perucká
Botič
Horská
Bělehradská
Botič
Sekaninova
Nuselský most
Vršovická
Otakarova
Botič
Ostrčilovo nám.
Ukrajinská
Jarimírova
Křesomyslova
Sezimova
Leningradská
VYŠEHRAD
Nuselská
Bartoškova
Čiklova
Culture Palace
Čiklova
V pevnosti
Táborská
VYŠEHRAD

HRADČANY CASTLE

The silhouette of the castle is perhaps the best-known view of Prague. With the advantage of its exposed position, the castle dominates the skyline of the left bank of the Vltava. Especially when floodlit at night, the broad front with the cathedral in the background is most impressive.

The historical importance of this royal residence matches its imposing appearance. Its history is not only tied up with that of the city, but also with the history of the first independent Czech state and its destiny. A thousand years ago the fate of the country was decided here, and this tradition has continued, with few interruptions, up to the present day. The castle is the seat of the President of the Republic, and still a centre of political power. The building of the castle dates from the same period as the first historically documented prince of the Přemyslid dynasty. Prince Bořivoj built what was at first a wooden fort on the site of a pagan place of worship. It became the seat of the dynasty and secured the crossroads of important European trade routes which met at the ford of the Vltava. At the same time, Bořivoj built the first church on the hill to replace the pre-Christian burial ground, as a sign of progressive Christianisation. In 973, when the bishopric of Prague was founded, the castle also became the bishop's seat.

After the turn of the millennium a Romanesque castle gradually evolved, with a princely (later royal) palace, a bishop's palace, several churches, two monasteries and massive fortifications. Every subsequent period has added its contribution to the complex development of the castle, of which you can get a rough idea when you visit. The castle as we see it today is mostly due to Empress Maria Theresa. In the second half of the 18th century, she commissioned the Viennese court architect Nicolo Pacassi to give the various buildings a unified, neoclassical facade and extensions. As a result, the individual character of the castle has been transformed into something more resembling a massive palace.

Hradčany Square: Before you start on your tour of the castle, take a look at the Hradčany Square (*Hradčanské nám.*) A few interesting palaces have been built since the destructive fire of 1541, which destroyed all of Hradčany and much of the Malá Strana lying below. Right next door to the castle you can see the rococo facade of the **Archbishop's Palace**, which is open to visitors only once a year, on Maundy Thursday. The Renaissance **Palais Schwarzenberg** on the opposite side of the square has painted sgraffito decoration which follows Italian models. In here is the **Museum of Military History** *(Vojenské muzeum)*

Preceding pages: Wenceslas Square. **Left**, entrance to the presidential palace.

with its unparalleled collection of weapons, uniforms, medals, flags and battle plans from all over Europe.

The bold proportions of its front facade draw attention to the early baroque **Palais Toscana**, which closes the square in the west. Where the Kanovnická ul. comes in you will see the Renaissance **Palais Martinic**. When the building was restored, sgraffito portraying biblical and classical scenes were discovered.

To the left of the Archbishop's Palace a little alley leads off to the hidden **Palais Sternberg**, which maintains its baroque splendour inside. This is the main building of the National Gallery (*Národní galerie*), which houses a first-class collection of European art.

On its south side, Hradčany Square opens up to the ramp leading up to the castle, from which you can get a superb view (as you can from the terrace of the *Café Kajetánka*). On weekends in the summer months you can get into the **Castle Gardens** through the entrance beside the **New Castle Steps**.

First and Second Courtyards: The main entrance to the castle complex is the **First Courtyard**, which opens onto Hradčany Square. You enter this so-called **Ceremonial Courtyard** through a gate in an immense ornamental wrought-iron railing. The guard of honour is posted in front of the statues, copies of the *Battling Titans* by Ignatius Platzer the Elder. The guard is changed every hour on the hour, a ceremony that always attracts a small crowd of onlookers.

This is the most recent of the courtyards and was built on the site of the western castle moat during the alterations of Maria Theresa's reign. Only the **Matthias Gate** is considerably older; indeed, it is the oldest baroque building in Hradčany Castle. It originally stood separate, like a triumphal

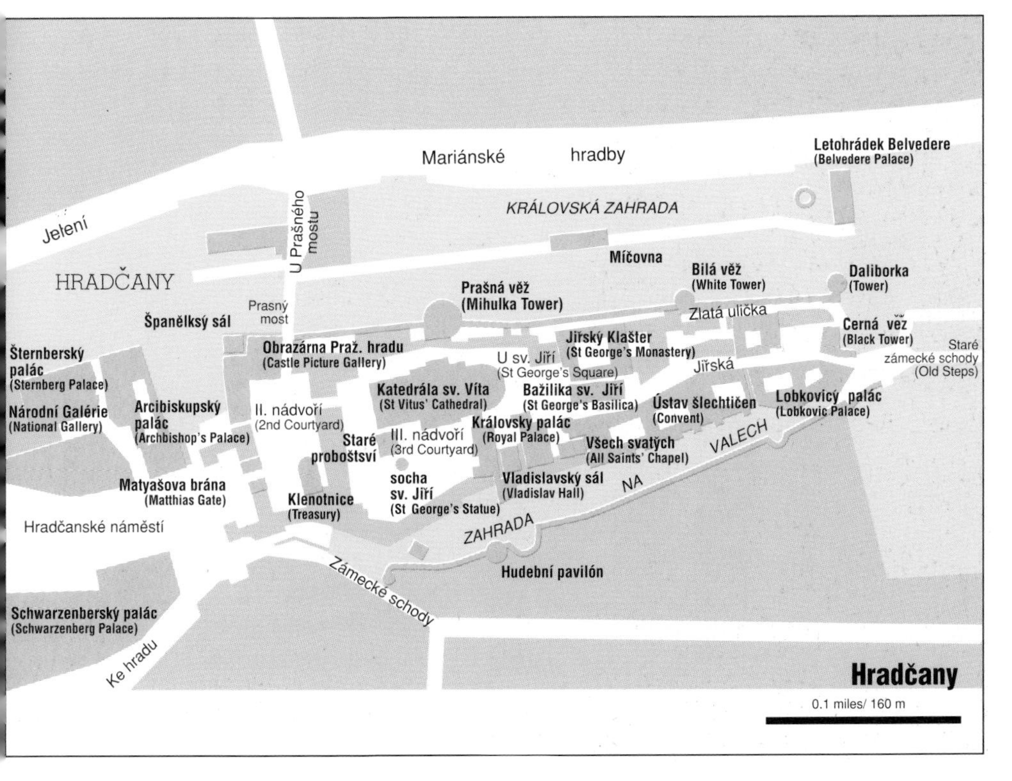

arch, between the bridges that led over the moats. During the rebuilding it was elegantly integrated into the new section as a relief. Since then, the Matthias Gate has been the entrance to the Second Courtyard. To the right in the arch of the gate a staircase leads off; this is the official entrance to the reception rooms of the President, which are only occasionally open to the public.

The **Second Courtyard** has a somewhat plain appearance. Once there, take a look first at the **Chapel of the Holy Cross** (*Kaple sv. Kříže*). The most valuable pieces of the **Cathedral Treasure** are kept in this former royal chapel with its magnificently decorated interior. They include a collection of valuable reliquaries, liturgical objects and interesting historical mementos. This fascinating collection originated in the days of Prince Wenceslas, but its core goes back to Charles IV. A great pragmatist in political matters, the Emperor was at the same time an impassioned collector of holy relics.

The symmetrical, closed impression given by the Second Courtyard also dates from Maria Theresa's innovations. However, behind this uniform facade lies a conglomeration of buildings which has grown up gradually over the centuries. Each has its own complicated history. In the right passage to the Third Courtyard you can see some remains of the Romanesque castle fortifications.

The remains of an even older building, the church of St Mary dating from the 9th century, were discovered in the **Castle Gallery**. Access to the gallery is from the passage in the north wing. Here you can see a collection largely put together by art lover Emperor Rudolf II. This emperor has gone down in history as something of an eccentric because of his esoteric way of life, yet he was a great patron of the arts and sciences and

The changing of the guard at the entrance to the First Courtyard.

collected a huge amount of art treasures, as well as countless curiosities. His unique collection was one of the most notable in the Europe of his day. When the imperial residence moved to Vienna, a great part of the collection went with it.

Still more fell to the Swedes as loot during the Thirty Years' War. Yet another valuable collection was created, still in the 16th century, from the remains, but much of it was taken to Vienna or sold to Dresden. What was left was auctioned off, and was thought for a long time to be totally lost. Only in recent years were pictures discovered during rebuilding work. These were restored and then identified as original paintings which had been believed lost. This small but valuable collection contains 70 paintings (among others, works by Hans von Aachen, Titian, Tinteretto, Veronese, Rubens, M. B. Braun, Adriaen de Vries and the Bohemian baroque artists J. Kupecký and J. P. Brandl).

The banner of the president flies over Prague Castle.

St Vitus' Cathedral: Once you walk through the passage and onto the **Third Courtyard** you can hardly avoid stopping and letting your eye follow the daring vertical lines of **St Vitus' Cathedral** (*Chrám sv. Víta*); the towering north portal is only a few steps away. The cathedral, the biggest church in Prague, is the metropolitan church of the Archdiocese of Prague, the royal and imperial burial church and also the place where the royal regalia are proudly kept.

The unique 600-year history of the building of the cathedral began when the archbishopric was founded in 1344. Ambitious as ever, Charles IV used this opportunity to begin the building of a cathedral which was intended to be among the most important works of the 14th century Gothic that was spreading from France. To this end Charles employed the French architect Matthias of Arras, who had trained in the French Gothic school and was working in Avignon (a papal city during those years). Arras died after eight years and the work was taken over by Peter Parler, who influenced all later Gothic architecture in Prague.

After his death, his sons continued the work, giving the building their own individual stamp, until it was interrupted in the first half of the 15th century by the Hussite Wars. It was in this period that the choir with its chapels and part of the south tower were completed.

Only a few small alterations were made during the years that followed. For instance, the tower was given a Renaissance top some time after 1560. A good 200 years later this was replaced by a baroque roof.

The difficult task of completing the cathedral was not attempted until the early 1860s, when a Czech patriotic association took it up. Following old plans and consulting famous Czech art-

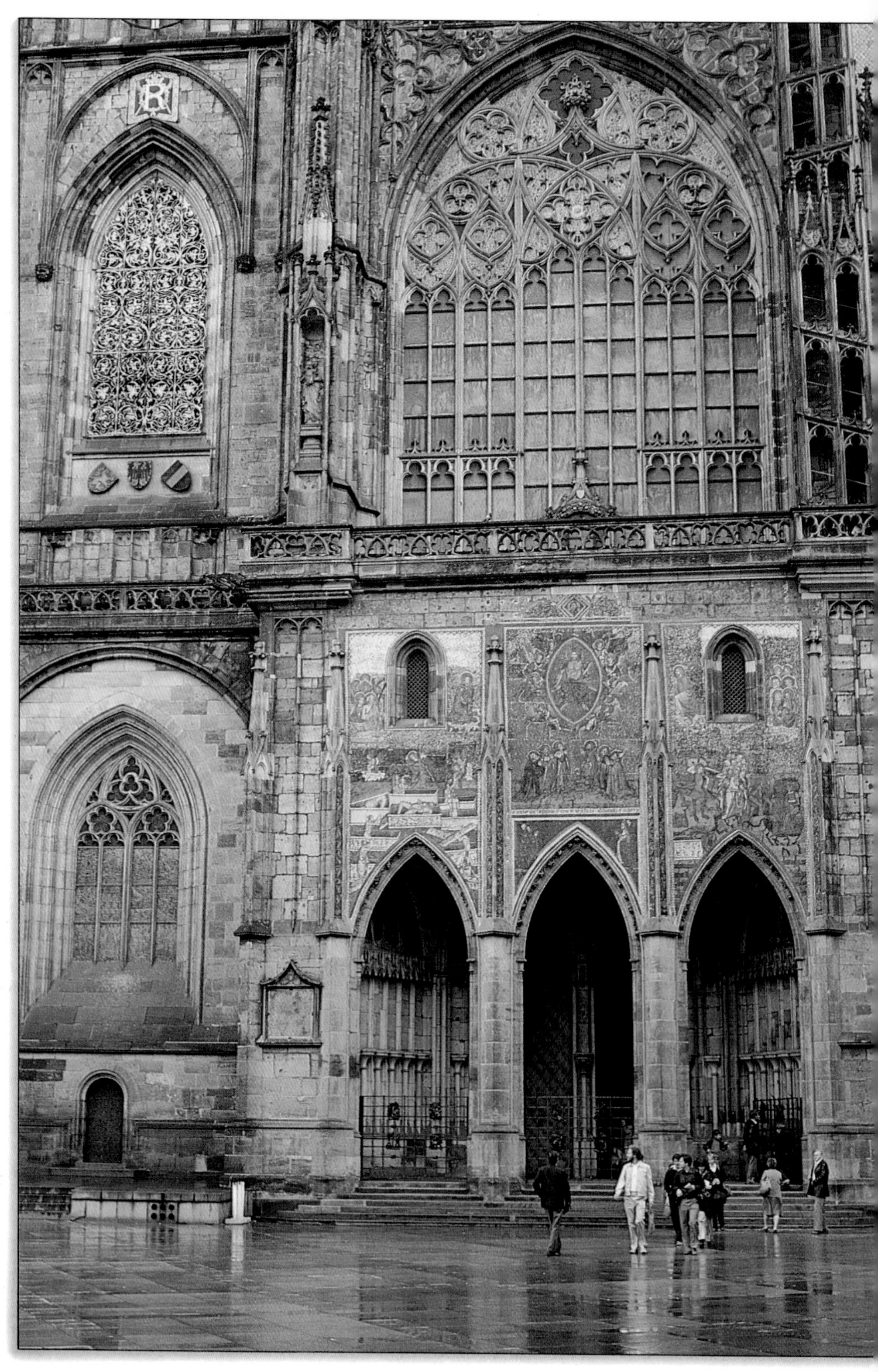

ists, they completed the building in 1929. All the extensions and additions carried out across the centuries explain why cathedral appears to lack a certain unity of style.

Before entering the cathedral through the western portal, take a look at its exterior, which dates from the last few years of the completion process. A notable feature is the **Rose Window**, more than 30 feet (10 metres) in diameter, which portrays the creation of the world. On either side of the window are portraits of the cathedral architects. The towers are decorated with the statues of the 14 saints. In the centre of the bronze gates the history of the building has been portrayed, on the sides you can see the legends of St Adalbert and St Wenceslas.

In the splendid interior of the cathedral the most notable features are the stained-glass windows and the triforium, a walkway above the pillars with a gallery of portrait busts. Leading Czech artists took part in creating the windows, among them Max Svabinsky, who was responsible for the window in the first chapel on the right, the mosaic on the west wall and the great window above the south portal. The window on the third chapel on the left was designed by Alphonse Mucha, who is perhaps best known outside Czechoslovakia for his Art Nouveau posters for Sarah Bernhardt. If you want to study all 21 chapels (each one contains several works of art), you should join a guided tour. Here we can only draw your attention to the most important sights.

St Wenceslas' Chapel: The main attraction is bound to be the **Chapel of St Wenceslas** which protrudes into the south transept. It was built by Peter Parler on the site of a Romanesque rotunda of the 10th century, in which the national saint Wenceslas was interred. In keeping with the importance of the Wenceslas cult, the saint's sacred place is exceptionally splendid in its ornamentation. The frescos on the walls, which are decorated with semi-precious stones and gold bezants, portray (in the upper half) Christ's passion and (in the lower) the story of St Wenceslas, prince of Bohemia. A little door leads to the **Treasure Chamber** directly above the chapel. Here the Bohemian royal regalia are kept, behind seven locks, the seven keys of which are kept by seven separate institutions. However, the precious jewels are only put on display on special occasions.

The three central chapels of the choir, behind the main altar, contain the Gothic tombs of the princes and kings of the Přemyslid dynasty. They are the work of Peter Parler's masons. In the choir itself, on the one side, is a kneeling bronze statue of Cardinal von Schwarzenberg (the work of the leading Czech sculptor J.U. Myslbek, 1848–1922), and on the other side is the massive silver tomb of St John Nepomuk, designed by the notable baroque architect Johann Emanuel Fischer von Erlach. Also remarkable are the wooden reliefs in the choir, masterpieces of baroque woodcarving.

Opposite the tomb of Count Schlick, designed by Matthias B. Braun, a staircase leads down to the **Royal Crypt**. Here you can see the remains of the walls of two Romanesque churches, and also the sarcophagi of Charles IV, his children and his four wives, George of Poděbrady and other rulers. The Emperor Rudolf II lies in a Renaissance pewter coffin. Above the Royal Crypt – just in front of the Neo-gothic High Altar – is the imperial tomb of the Habsburgs, built of white marble for Ferdinand I, his wife Anna and their son Maximilian.

Visitors should not forget to glance upwards to admire the lozenges adorning the roof of the choir. Peter Parler displayed a masterly ability to combine revolutionary technical solutions with elegant caprice. This is particularly evi-

Left, imposing and overpowering: St. Vitus' Cathedral in Prague Castle.

dent on the south side of the choir, where the interplay of columns and struts and the remarkable complexity of the tracery are especially impressive. The organ loft originally marked the end of the choir on the west side. Once the Neo-gothic part of the cathedral was completed, it was moved to its present position.

Third Courtyard: In order to see the rest of the interesting sights, you have to walk around the former **Old Chapter House**, (nowadays the House of Culture) which is pressed up against the side of the cathedral. Of special interest to art historians is the equestrian statue of St George which stands prominently in the courtyard. It is, however, a copy of a Gothic sculpture. The original is in the St George monastery and is evidence of the highly developed art of 14th-century metal casting. The flat-roofed shelters next to the cathedral are to protect archaeological discoveries made in the lower levels of the castle courtyard.

From here you can get an impressive view of the complex system of buttresses and the south facade of the cathedral, which is dominated by the 300-ft (nearly 100-metre) tower. Its stylistically unusual top gives it an individual appearance. The gilded window grille, the letter "R" and the two clocks (the upper shows the hours, the lower the quarter hours) date from the time of Rudolf II. Unfortunately, it is no longer possible to climb the tower. It contains four Renaissance bells, among them the biggest bell in Bohemia, which weighs 18 tonnes.

Unusual in both position and execution is the **Golden Door**, the distinctive portal that leads into the south transept. It is the main entrance to the cathedral; it was through here that monarchs passed on their way to their coronation. Its remarkable triple-arched anteroom

Detail of the Rose Window of St Vitus' Cathedral. It portrays the Creation.

has an exterior mosaic depicting *The Day of Judgment.* It was created by Italian artists around the year 1370. The anteroom is fitted with a grille depicting the individual months.

The covered staircase in the left-hand corner leads to the castle gardens, mentioned at the beginning of this chapter. These lead to the former **Royal Palace** (*Královský palác*) which should definitely be seen. It should no longer be a surprise that this complex was also built by many different generations of rulers. New storeys of the palace were layered one above the other on top of the oldest walls, which now lie deep under the level of the courtyard. Now you can, in a manner of speaking, literally gain deeper and deeper insights into the castle's past.

A Royal Tour: Go past a fountain featuring an eagle, and from the courtyard you will be able to ascend the staircase leading to the anteroom. From here, you can start your tour of this palace, which up until the 16th century was the residence of the rulers. The first three rooms to the left of the entrance constitute the **Green Chamber**, a former law court and audience hall (it has a ceiling fresco, *The Judgement of Solomon*). Further along is the so-called Vladislav Bedchamber and the Land Records Depository. The Land Records were books in which not only the details of property ownership but also the decisions of the Bohemian Estates and of the law courts were recorded.

Leave the anteroom and go on to the **Hall of Vladislav**, named after King Vladislav II. This unique, most imposing Late Gothic throne room was built by the architect Benedikt Rieth between 1493 and 1502. Numerous coronations and tournments took place under the 43-ft (13-metre) high pillars. Immediately to the right of the entrance of this hall another wing of the building

A few years ago the roof of St Vitus' Cathedral was still open to the public.

is joined. Continue on the same level and you will come to the **Bohemian Chancellery**. In the first room is an impressively clear model which shows the appearance of the castle in the 18th century and compares it to today. Go through a Renaissance portal and you will enter the actual office of the former imperial governor.

This room became famous as the scene of the so-called "Second Defenestration of Prague", which marked the beginning not only of the Bohemian rebellion but also of the Thirty Years' War. On 23 May 1618, two Catholic governors and a secretary were thrown out of the left-hand window because they had broken the terms of Rudolf II's *Majestát*, or "Letter of Majesty". A few years previously, the emperor had guaranteed the Bohemian nobility freedom of religion with this decree. Two obelisks in the garden mark the spots where the two honorable gentlemen are supposed to have landed. No such honour was accorded on the secretary. All three survived the defenestration practically unscathed, for they allegedly fell into a dunghill in the castle moat.

Go up a spiral staircase and you will come to the **Imperial Court Chancellery** which is situated above the Bohemian chancellery. Under Rudolf II's rule, the whole Holy Roman Empire was ruled from here.

Saints and Fortifications: Under the three Renaissance windows on the narrow wall of the Vladislav Hall a staircase leads off to the **All Saints Chapel**, which contains three remarkable works of art: the *Triptych of the Angels* by Hans von Aachen, the painting of *All Saints* on the high altar by V.V. Reiner and, in the choir, a cycle of paintings by Dittmann. The latter portray 12 scenes from the life of St Procopius, who is buried in the chapel.

St George's Fountain in the Second Courtyard.

The next room leading off from the Vladislav Hall is the **Council Chamber**, in which the Bohemian Estates and the highest law court assembled. The royal throne and the furnishings are 19th century. To the left of the throne is the tribunal of the chief court recorder, built in a Renaissance style. The wall is decorated by the portraits of the Habsburg rulers.

The last room open to the public in this wing is the **New Land Records Office**, with the heraldic emblems of the Land Records officials on the ceiling and along the walls.

The **Riders' Staircase**, which was constructed so that it should be wide enough to for rulers and guests to enter and take part in the riotous festivities on horseback, leads out of the most recent part of the palace. If the lower storeys are accessible you can continue your tour to the left, going down into the early Gothic levels of the palace. The lowest level is the Romanesque palace which contains the remains of fortifications partly dating from as early as the end of the 9th century.

Skilful wrought ironwork in St Vitus' Cathedral.

Leave the Royal Palace and go out into St George's Square (*Nám. U sv. Jiří*). The baroque facade opposite the end of the choir of St Vitus' Cathedral belongs to the **Basilica of St George** (*Bazilika sv. Jiří*). This is the oldest church still extant on the site of the castle. Together with the adjoining monastery, it formed the hub of the castle complex in the early Middle Ages. It was founded in the early 10th century and by the 12th century the Romanesque church as we know it today had been completed. Despite extensions and rebuilding programmes during the Renaissance and baroque periods, the church has retained its early medieval appearance and, following major renovations carried out at the beginning of the 20th century, has been restored to its former glory.

The beautiful interior, in which concerts are held because of the excellent acoustics, is closed off by a raised choir. Here you can still see remnants of the original Romanesque ceiling paintings. To the right of the choir you can look through a grille into the **Ludmilla Chapel**, with the tomb of the saint, the grandmother of Prince Wenceslas. The tombs of two Bohemian nobles are placed in front of the choir. The baroque statue in front of the crypt – a corpse with snakes in its intestines – is a realistically portrayed allegory of the transitory nature of life.

The baroque **Chapel of St John Nepomuk** is incorporated into the outer facade of the basilica. Its portal is decorated with a statue of the saint by F.M. Brokoff.

Adjoining the basilica on the left is the former Benedictine **Monastery of St George** (founded 973), rebuilt several times, which today houses the **Old Bohemian Art Collection** (part of the

National Gallery). On exhibition here are works from the 14th to the 18th centuries, among them pictures by artists who took part in the building of many of Prague's churches.

Along the north side of St Vitus' Cathedral runs the Vikárská ul., in which there is nowadays a tourist information office. Nearby is the **Milhulka Powder Tower**, which in recent years has been opened to the public. In the late 15th century, while parts of the northern fortifications were being built, it served as a gunpowder magazine. Today it contains a small museum, which records the tower's earlier use as a metal casting foundry and possibly an alchemist's laboratory. Individual storeys portray aspects of crafts in the 16th and 17th centuries.

Goldmakers' Alley: Another part of the fortifications can be seen behind St George's monastery. Go past the basilica and turn left up the **Goldmakers' Alley** (*Zlatá ulic ká*), considered to be one of the most popular attractions of the castle. In the part of the fortifications between the central **White Tower** and the outermost **Daliborka Tower**, tiny houses crouch under the walkway on the wall, making a romantic backdrop. Legend has it that this is where the famous alchemists employed by Rudolf II tried to discover the secret of eternal life and how to make gold. It is a fact, however, that the author Franz Kafka occasionally lived and worked on his novels in house no. 22.

The tour of the castle ends at the Black Tower, where the Jiřská ul. reaches the eastern gate. Just before the gate, on the right-hand side, you come to the **Palais Lobkowic**, which in recent years has been used for many exhibitions based on themes from the country's history. On the other side of the eastern gate the **Old Castle Steps** and the street Na Opyši lead down to the Malá Strana and the Metro station, Malostranská.

The Belvedere Palace: Lying outside the castle complex, the Belvedere Palace (*Královský leto-hrádek*) is worth visiting. Leave the Second Courtyard and go north, cross a bridge (*Prašný most*) going towards the former **Riding School** (*Jízdarna*), nowadays an exhibition hall. From its terrace you can see an imposing view of the St Vitus' Cathedral. Leaving the Riding School, follow the **St Mary's Wall** (*Mariánské hradby*), which borders the Royal Gardens (unfortunately closed to the public) round to your right. It is not far now to the splendid palace, which art historians consider to be the only example of a purely Italian Renaissance building north of the Alps. The Emperor Ferdinand I had it built in the mid-16th century for his wife Anna. Especially remarkable is the "singing" Renaissance fountain in the garden, which was cast with such skill that the falling water, striking the bronze basin decorated with hunting scenes, makes it ring.

Left, Goldmaker's Alley reflected in the water of the fountain. **Right**, wild vines on the Old Castle Steps.

THE GOLEM – A MYTH IS BORN

Every city with a long history has its stories, in which real events are shrouded in mysterious shadow. Prague has such legends even about its founder, the Princess Libuše. There is also the legend of the headless Templar knight who rides through Prague, which dates from the persecution of the Templar order. And there are the water spirits who guard the many dams of the Vltava. There is also Doctor Faustus, who tried to make gold, using a pact made with the devil, no less. Finally the Prince of Darkness dragged him off to his kingdom forever, drawing him through the ceiling of the Faustus house. This legend is probably based on the figure of the alchemist Mladota, who lived in Prague.

However, none of these tales ever obscured the real, historical Prague. Not until the late 16th century did an image of a fantasy Prague begin to form. These were strange times. The glory of the Renaissance was giving way to the twilight of the baroque. The Renaissance had freed humanity of the superstition of medieval times, but was not yet able to force its discovery of reason on the world. Certainty seemed to have been reached, but then new doubts, new uncertainties arose. The new knowledge blundered on into the uncertainties – searching, believing, doubting, mistaking, often falling into despair.

And yet the Golem, a fantastical-mechanical monster, an animated piece of mere matter, could really only have come into existence in the mysterious ghetto of the time of Rudolf II. It was this Roman Catholic emperor and Bohemian king in the midst of a country of heretics who created the image of this fantasy Prague, a city of alchemists and artists, of astrologers and scholars, who tried to lift the veil of divine secrets. A world of dreamers, of deluded seekers for truth and of lost pilgrims in search of eternal verities. This was the atmosphere of the city in which the miracle-working Rabbi Löw and his Golem could come face to face with the most famous magician of all, Doctor Faustus. A fantasy Prague, whose atmosphere of magic and legend cannot be equalled by any other epoch. Especially not by our own times, when scientific discoveries have long usurped all the mysteries of the sages of Rudolf's era.

Here then is the legend of the Golem, the creature of mud and clay made by the cabalist, astronomer and magician Rabbi Löw, who breathed life into the Golem with a magic word, the "Shem", in order, as the *Sippurim*, a 19th-century collection of Jewish legends, relates, "to send it out to protect his community, to discover crimes and to prevent them". One evening, before the Sabbath rest, Rabbi Löw forgot to remove the sign of life from the mouth of the Golem, and the latter began a rampage of destruction in Löw's house. The spark of life was removed and the creature turned back into mud and clay, to lie forever under the roof of the Old New Synagogue.

The origins of the legend itself are based on the first part of the *cabala*, the mystical teachings and writings which are already mentioned in the Talmud, but which contain no word of an artificially created human being. Not until the commentary of Eliezer of Worms, dating from the 13th century, does the word "Golem" in the sense of an artificial creation appear (along with exact instructions for making such a creature). In the medieval stories, the Golem is portrayed as a perfect servant, its only fault being that it interprets its master's instructions too literally. By the 16th century, it was seen as a figure that protected the Jews from persecution, but it had also acquired a sinister aspect. Not until the middle of the 19th century do we find any connection in writing between these tales of the "creative" Rabbi and the figure of Rabbi Löw the alchemist. According to this version, Rabbi Löw, clothed in white, went one dark night to the banks of the Vltava and there, with the help of his son-in-law, he created the Golem, while continually chanting spells, from the four natural elements – earth, fire, air and water.

This story forms the basis of the German novel *Der Golem* by Gustav Meyrinck. In the 1920 it was made into a film, which became a classic of German silent cinema, and the Golem became the model for many other cinematic man-made monsters, including Frankenstein.

Left, tombstones in the old Jewish cemetery. Above, the Star of David, symbol of the Jewish faith, located above the Jewish Town Hall in Maiselova.

THE STRAHOV MONASTERY

It lies outside the castle fortifications and outside the whole castle complex, on the age-old trade route from Nuremberg to Cracow, on the slopes of the Petřín Hill, the crown of the gently sloping valley. It is the **Strahov monastery**, the oldest Premonstratensian monastery in all Bohemia, now lying on the Strahovské nám. The two towers of the Strahov, along with the green of the PetřínHill and its miniature "Eiffel" tower, and the long line of the roof of the Palais Czernin, all make up the unmistakable silhouette of the left bank of the Vltava.

The first monastery of the white monks of the Premonstratensian order was founded in 1140 by King Vladislav II, but was completely destroyed by fire in 1258. The wars of the ensuing centuries also left their mark, with the result that very little remained of the original Romanesque building. Today the monastery is predominantly baroque in style, but it contains early Gothic and Renaissance elements. Only St Mary's church retains traces of the Romaneque original.

The monastery continued to function until 1952. After the dissolution of all religious orders in Czechoslovakia, it was declared a museum of national literature and opened to the public in this guise on 8 May 1953. This rapid transformation was possible because of the vast resources of the monastery library, which had been slowly gathered over the centuries. Today, the monastery possesses not only the oldest and most extensive, but above all the most valuable library in the country. The core of the collection was established at the time of the foundation 800 years ago. Gradually added over the years were examples of almost the complete literature of western Christianity up to the end of the 18th century. Today the emphasis is on national literature of the 19th and 20th centuries.

If you go into the harmonious enclosure of the monastery, the first thing you will see is the church of St Rochus, built from 1603 to 1612 during the rule of Emperor Rudolf II, and nowadays used as a gallery. On the facade of the "New" Library (built from 1782 to 1784) is a medallion with the portrait of Emperor Joseph II, the ruler whose support of the Enlightenment led to the dissolution of the majority of monasteries in his domains. His memory is honoured here because he allowed Strahov to remain, and the monks of Strahov to buy the complete inventory for the building of a new library from another famous monastery in Moravia, the Bruck monastery near Znojmo. These brown and gold gleaming shelves equipped the new hall, which was then designated the "Philosophers' Hall". The older hall was renamed the "Theologians' Hall".

Strahov Library: The **Theologians' Hall** was built by Giovanni Domenico Orsi in a rich baroque style at a cost of 2,254 guilders. It was painted with splendid ceiling frescos by Siardus Nosecký, a member of the monastery, from 1723 to 1727. The theme is true wisdom, rooted deeply in the knowledge of God. The brightly coloured scenes in their sturdy stucco frames radiate warmth and cheerfulness. In the middle of the room stand a number of valuable globes from the Netherlands, dating from the 17th century.

The ceiling fresco in the **Philosophers' Hall** is less easy to read. The work of Anton Maulbertsch, it is in concept and technique a monumental finale to rococco ceiling painting in Europe. The fresco shows the development of humanity through wisdom – a theme that borders on the ideas of the Enlightenment. On the two narrow sides you can see Moses with the tablets of law and, opposite, Paul preaching at

Preceding pages: the Theologians' Hall in the library of the Strahov monastery.

the pagan altar. The long lines of figures on the long sides introduce the great personalities of history who have made tremendous progress possible through their achievements.

The Strahov Gospels: In 1950, the library contained 130,000 books. By now this number has grown to around 900,000, as the Strahov has taken in the contents of a number of other monasterial libraries, particularly in central and northern Bohemia. One of the most famous of all manuscripts is the Strahov Gospels, the oldest manuscript in the library, dating from the 9th to 10th century during the reign of Charles IV. A copy can be seen in the Strahov. Also among the most valuable treasures are rarities such as the New Testament, printed in Plzeň in 1476 and one of the first printed works in Czech, or the beautifully illustrated story of the journey of Frederick von Dohna to Rome, dating from the 17th century. A great deal of this can be seen in the photographs and transparencies that form part of a special cultural and historical tour set up in the cloisters. A separate room is devoted to the great reformer, Jan Hus. The upper storey contains a great number of documents relating to the writers who influenced the re-awakening of Czech national consciousness in the 19th century.

Church and Monastery Gardens: If the church happens to be open, it is best not to miss the opportunity of visiting and admiring this originally Romanesque building, which was vastly altered and richly redecorated in the 17th and 18th centuries.

Also forming part of the monastery grounds are the large gardens. These fill the valley between the Petřín and the castle hills right up to the edge of the Malá Strana. Once white-robed monks walked here; nowadays it is a favourite haunt for courting couples.

'iew over ie rooftops f the Malá trana.

THE VLTAVA

The Vltava (Moldau) is 270 miles (435 km) long and rises from two headstreams – the Teplá Vltava, rising on the mountain of Černá hora, and the Studená Vltava. It flows first south-east, then north across Bohemia. Once a wild river whose floods threatened the city, the Vltava today flows peacefully and calmly. Dams – among them three large hydrodams with lakes that are also used for recreation – raise the water level and slow down the current, and the Vltava gives an impression of might and majesty. When it

reaches the town of Mělník, it joins the Elbe, called Labe in Czech, (the latter only becomes navigable from this point on), and by this route connects Prague with the West German seaport of Hamburg. Thus the Vltava forms an important link for river traffic between Czechoslovakia and the North Sea. The **Prague river harbour** lies in the bend of the Vltava in the city district of Holešovice. Nothing remains of the earlier harbours which once lay on a tributary (now filled in) of the Vltava except the pub "Hamburk" in the square of Karlínské nám.

Apart from freight and shipping, the river is mostly used for recreation and enjoyment. Boats for river excursions leave from the Palacký Bridge. They travel in both directions, downstream to the zoo and the Prague suburb of Roztoky, upstream to the artificial lake of Slapy. Rowing boats can be hired and are very popular – it is impossible to imagine the Vltava in summer without them. The winter fairs on the frozen river are unfortunately a thing of the past, as the damming of the river has warmed the water and the river no longer freezes. The oldest open-air swimming pool on the Vltava dates from 1840 and is situated on the Malá Strana bank next to the S. Čech Bridge.

From the hills of the left bank you can get a beautiful view of the arrangement of the bridges, which lie one behind the other. For 500 years the old Charles Bridge was the only link between the two river banks. Not until the mid-19th century Industrial Revolution were more bridges built. At about this time, a start was made on building the embankments. Only the Malá Strana bank in the vicinity of the Charles Bridge has been altered in this way. Kampa Island and the mouth of the Čertovka tributary have remained in their original state.

The Vltava has been honoured in various different ways. The allegorical representation of the Vltava is popular – it's a statue which decorates a fountain on the exterior of the garden wall of the Clam-Gallas Palais. Its popular name is **Terézka** – rumour has it that a wealthy citizen of Prague left his whole fortune to the statue. In the late 18th century, at the time of the birth of the Czech nationalist movement, the Vltava was a never-failing source of inspiration. The Vyšehrad castle on its rock above the Vltava was famed in many legends, and in the Romantic period it fascinated many Czech artists searching for their national identity. The myths and the river that was linked with them found their expression in countless songs, works of representational art and literature. Later, this tradition was continued by a whole generation of artists who took part in the construction of the National Theatre, that symbol of the completion of national rebirth. The building was, appropriately, sited on the bank of the Vltava, not far from the Vyšehrad rock, where it forms another dominant feature of the city skyline. The theatre was opened in 1881 with a performance of Smetana's *Libuše*, in which the solemn prophesy of the mythological princess is heard. According to another legend, mostly omitted from the artistic versions, Libuše had a very prosaic relationship with the river. It is claimed that, when she had enough of her lovers, she had them thrown from the Vyšehrad rock into the Vltava. The national composer Smetana took up the Vyšehrad myth once again when he composed the symphonic poem *Má Vlast* (My Fatherland). The second movement deals with the Vltava and is perhaps the most famous artistic representation of the river. Even today, the river still influences people's imagination, though in a quite different way. Children in particular are familiar with the Vltava water spirits which appear in fairy tales. The Czech water spirits, little men with green coats and pipes, have lived in the water since time immemorial. The water spirits know every stone and every fish, are very wise and are always ready to give good advice.

Preceding pages: Charles Bridge. Left, once upon a time – winter amusements on the frozen Vltava. Above, spring flooding.

MALA STRANA

The Malá Strana (Lesser Quarter) lies at the foot of the castle of Prague. It is a totally individual quarter, a picturesque island, separated from the rest of the city by broad parks and the wide, steady flow of the Vltava. Looking down from the hills, the impression gained is of a landslide of roofs which started to roll between the Hradčany and Petřín hills and came to a stop on the river bank.

In 1257, the Malá Strana was made a city, and is thus the second oldest among the four historic cities that make up Prague. The Malá Strana experienced its first boom during the rule of Charles IV. During this time it was extended considerably and received new fortifications. However, not until catastrophic damage was inflicted by the great fire of 1541 was there any sign of a major rebuilding programme. The rebuilding after the fire shaped the individual characteristics of this quarter and we can still see them today.

The Malá Strana truly blossomed after the victory of the Catholic League over the Bohemians in the battle of the White Mountain in 1620, when many wealthy families loyal to the House of Habsburg settled here. True, most of the palaces were deserted once the political administration of Bohemia had moved to Vienna, but the palaces have been spared major alteration to this day. Even the town houses, which often have much older foundations, have kept their mainly baroque facades with their characteristic house signs. For this particular reason, the Malá Strana can be described as an architectural jewel, indeed as a complete work of art representing the baroque style of Central Europe. The different creative styles of the town houses, the small, quiet squares, the

View of St Nicholas and the roofs of the Malá Strana.

mansions with their gardens designed to blend in with the hill slopes all came together to form an original style – "Prague baroque".

Of course, many things worth seeing remain hidden behind the facades. However, once off the main streets, you can enjoy the special atmosphere of the place. This is not least due to the fact that by no means everything revolves around the tourist, and the Malá Strana has its own everyday life to live.

Around Malostranské náměstí: The centre of the Malá Strana always was and still is the Malostranské námestí, a square which is actually divided into two squares by the **Church of St Nicholas** (*Chrám sv. Mikuláše*) and the neighbouring former Jesuit college.

The conspicuous dome of the church of St Nicholas and its slender tower can be seen from many different viewpoints in an ever-changing perspective. This unequal couple have become the symbol of the whole Malá Strana. The church itself is a masterpiece of baroque architecture and one of the most beautiful examples of its kind. In the early 18th century the famous Bavarian architect Christoph Dientzenhofer built the nave and side chapels on the site of a Gothic church. The choir and the dome were added later by his son Kilian Ignaz. The building was completed in the mid 18th century by the addition of the tower, the work of Lurago.

Particularly outstanding among the special features of the interior is the monumental ceiling fresco by J.L. Kraker in the nave. It is one of the biggest in Europe and portrays scenes from the life of St Nicholas. Another valuable fresco, by F.X. Palko, decorates the dome. The dome is 247 feet (75 metres) high – tall enough to accommodate the tower on the Petřín Hill inside it. The sculptures in the choir and

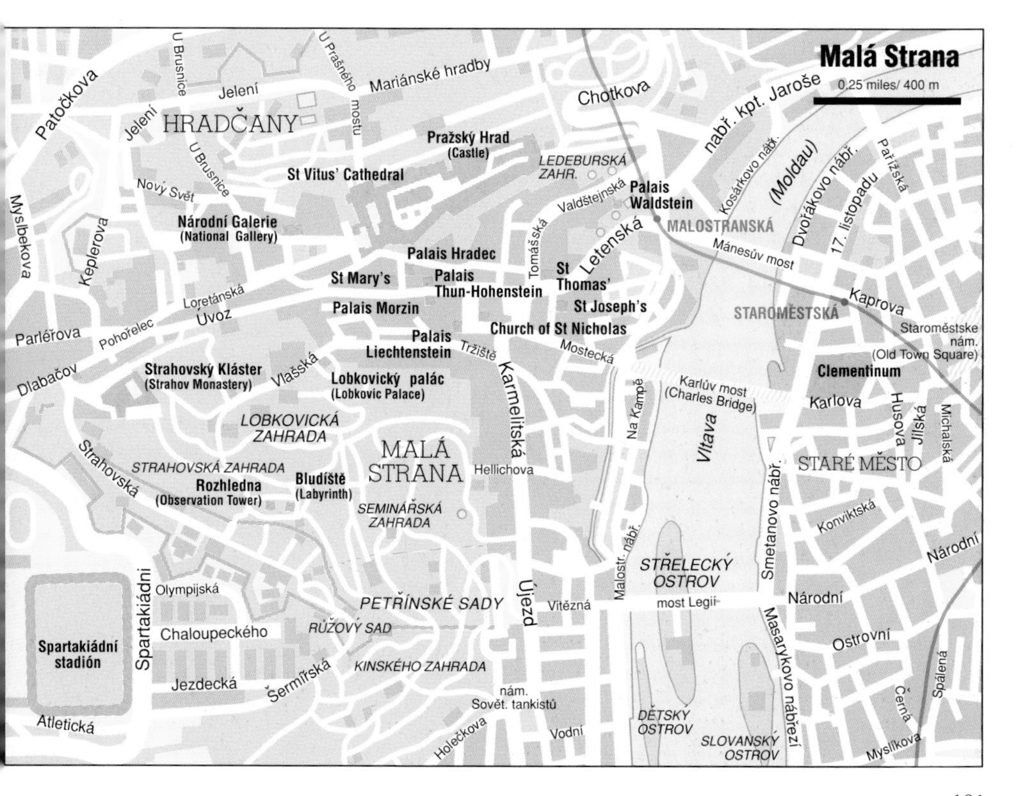

VYSKA
VRCHNÍHO
VEDENÍ
3,95m
22

the gilded statue of St Nicholas, patron saint of merchants and sailors, are the work of Ignaz F. Platzer the Elder.

Opposite the church is the **Palais Lichtenstein** with its broad neoclassical facade. From 1620 to 1627 it belonged to Karl von Lichtenstein, the so-called "Bloody Governor" who was mainly responsible for the execution of the leaders of the 1618 rebellion.

From St Nicholas' Church you will notice the **Golden Lion** house (*U zlatého lva*). It is one of the few purely Renaissance houses in the Malá Strana and also contains the wine bar *U mecenáše*. These small wine bars, whose charm lie mainly in their ancient walls, are typical of the Malá Strana. Guests were served here as early as 1600. Nowadays, however, it's not always easy to find a seat. For beer drinkers, the pub **U Glaubicû** has always been a household word. The corner house of the same name, a little further on under the arcaded passage, is being renovated. We will have to wait and see if the tradition of excellent beer continues after the work is complete.

Cross the street **Karmelitská** and follow the arcaded passage past a back courtyard rich in atmosphere. It lies hidden behind the arch of a gateway. All over the quarter, little surprises such as this await the observant visitor. However, visitors should cultivate a sense of tact, because even the proverbial hospitality of the Czechs, which is under considerable strain during the tourist season, has its limits.

The lower side of the square is bordered on the right by the **Palais Kaiserstein**. A memorial plaque inside the house proclaims that the world-famous opera singer Emmy Destinn once lived here. On the left side is the former town hall of the Malá Strana (it is inscribed "Malostranská beseda"). Today it's the place to go for anyone who wants to hear jazz in Prague. The well-established **Café Malostranská kavárna** protrudes into the square and, in the summer months, provides a welcome opportunity for eating and drinking out of doors.

eft, trams
the
etenská.

Leave the lively Mostecká ul. and on your right, you will find that the streets have become quieter. Here you enter one of the quiet, dreamy corners of the Malá Strana. There is a passage next to the cinema *"U hradeb"*, or you can turn off into the Lázeňská ul.. House no. 6, *"Vláznich"* (The Spa) was a first class hotel in the 19th century. Amongst the celebrities to have stayed here was Tsar Peter the Great. A memorial plaque proclaims that the French poet René Chateaubriand was also a guest. Another memorial plaque, on the house "*U zlatého jednorožce*" (The Golden Unicorn), marks the place where Ludwig van Beethoven stayed.

You can find a beautiful baroque interior in the **Church of St Mary in Chains** (Kostel Panny Marie pod řetězem), the oldest church in the Malá Strana. Remains of the walls of its predecessor, a 12th-century Romanesque basilica, can still be seen in the right-hand wall of the forecourt.

An excursion to your right will take you to the long sprawl of the **Maltese Square** *(Maltézské nám.)*, with its sculpture representing John the Baptist, which is the work of F.M. Brokoff. Two palaces in this square are worth your attention: the **Palais Turba** (now the Japanese Embassy) with its rococo facade, and the early baroque (later rebuilt) **Palais Nostitz** (Dutch Embassy, Ministry of Culture) which closes off the square to the south.

Adjacent to St Mary's church is another little square – the Velkopřevorské nám. On the one side is the **Palais Buquoy**, home of the French Embassy, and opposite is the former **Palace of the Grand Prior of the Knights of Malta**, one of the most beautiful in the Malá Strana. Today it contains the **Museum of Musical Instruments**, and its vast collection, which will also appeal to non-musicians, is well worth seeing. In the summer, numerous open-air concerts are held in the adjoining "Maltese Gardens".

A little bridge connects the square with the island of **Kampa**. It is separated from the Malá Strana by a tributary of the Vltava, the *"Čertovka"*. During the last decade of Communist rule, the island and the palace gardens became a favourite meeting place for the "Flower Children of Prague". They left behind lovingly executed murals in the district of the city known as "Little Venice" on account of its nostalgic water mills and gardens. The park has been formed by linking up the gardens of former palaces and offers a beautiful view of the Old Town. Between the Charles Bridge and the mouth of the Čertovka there is a small group of houses, lying directly by the water, which are popularly known by the proud name of "the Venice of Prague".

Go up a double flight of steps and you will come up onto the **Charles Bridge** *(Karlův most)* which, together with the silhouette of the Hradčany castle, has become the symbol of Prague. A wooden bridge, linking the two banks of the Vltava at approximately the same place, is mentioned as early as the 10th century. In 1165 it was probably replaced by the stone Judith Bridge, the second oldest stone bridge in central Europe. After the destruction of the Judith Bridge in a flood, Emperor Charles IV had a new one built by his cathedral architect Peter Parler. Named after the emperor, this venerable bridge was also damaged by floodwater several times, but never collapsed. According to a persistent legend, eggs were mixed with the mortar to give it durability. Perhaps, also, it is no coincidence that the foundation stone was laid on 9 July, the day of Saturn's con-

Left, archer in the Waldstein gardens. Right, house sign "The Three Violins".

junction with the sun. In any case, the astrologers, who in those times were often consulted when important decisions were to be made, considered it to be a most auspicious moment. These and other theories have often been cited to try and discover the secret of the bridge. Whatever it may have been, this 600-year-old construction is deserving of admiration, especially as it even stood up to the car and tram traffic of the 20th century, until it was declared a pedestrian precinct and restored to its well-earned rest

Statues on the Bridge: The mainly baroque statues (nowadays partially replaced by copies) were created in the late 17th century after the model of the Ponte Sant'Angelo in Rome. In contrast to the Gothic architecture, these give the bridge its characteristic appearance. The famous artist Johann Brokoff, his sons Ferdinand Maximilian, Michael Josef, and Matthias B. Braun were among those who worked on the statues. The last named was responsible for what is perhaps artistically the most valuable sculpture: the **statue group of St Luitgard** (1710). It shows Christ appearing to the blind saint and allowing her to kiss his wounds.

The route from the Charles Bridge to Malá Strana Square crosses the Mostecká.

The oldest statue on the bridge is that of **St John Nepomuk** (1683), designed by M. Rauchmuller and Johann Brokoff. The reliefs around the base portray scenes from the life of the cleric, canonized in 1729. One relief is based on the legend of John Nepomuk's death: the wife of Wenceslas IV (one of the sons of Charles IV) had made her confession to Nepomuk, whereupon the king put pressure on him to disclose the details. John Nepomuk steadfastly refused to break the seal of the confessional, and in 1393 Wenceslas IV had him drowned not far from this spot. In actual fact his belated canonization probably had more to do with the intention to oust the memory of Jan Hus with the cult of a new saint.

Another popular figure is the **statue of Bruncvik** on one of the bridge supports on the banks of the island of Kampa. The sword of this legendary knight, who is associated with the Roland epic, is walled up – or so it's said – in the bridge and can be retrieved in the country's darkest hour.

At the end of the bridge are the **Malá Strana Bridge Towers**, with an archway in the middle. Among the coats of arms that decorate it is that of the Old Town. The bridge, including the Malá Strana towers, actually belonged to the Old Town. That is why the former customs house can be found on this side of the bridge, on the left in front of the gate. The smaller tower is a remnant of the Judith Bridge, and only its Renaissance gables and the wall ornaments were later added. The higher tower, dating from the 15th century, was designed to complement the Old Town tower. The top is open to the public. If you look down towards the Old Town, you will see why Prague is also known as the "city of a hundred spires".

Downstream from the Charles Bridge the palace gardens of the Malá Strana beckon. Some of them are open to the public in the summer months. Without leaving the bridge you can see the Renaissance house *U tří pš rosů* (The Three Ostriches). The remains of its sgraffito decoration show that it once belonged to a supplier of feathers to the royal house. This house points the way to the street U lužického semináře, where the former **monastery garden** *(Vojanovy sady)* is situated. In this park with its two baroque chapels modern sculptures are often exhibited.

You certainly should not miss the chance to visit the **Waldstein Garden**, which can be reached from the Letenská ul.. We shall refer to it again later in connection with the Palais Waldstein. If you not only know about Czech beer (the dark beer, in this case), but consider it to be an essential form of sustenance, you should visit the traditional inn **U sv. Tomáše** (St Thomas') in the same street. The beer garden of this former monastic brewery (founded 1358) is unfortunately no longer in existence, and neither is the home brew, but Braník beer tastes no less excellent down in the cellar vaults.

A church is, of course, also part of this former Augustinian monastery. It was founded in the 13th century. Its present baroque form is the work of Kilian Dientzenhofer, and inside it is richly decorated with the works of Bohemian baroque artists.

Malá Strana Gardens – Palais Waldstein: When making a tour of the Malá Strana gardens, you should also take a quick look at the Underground station *Malostranská*, which contains a copy of M.B. Braun's *Hope*. Some sculp-

Canoe training in a nearby stream.

tures from his workshop can be seen in the courtyard garden. Your route now takes you on through the Valdštejnská ul. past palaces whose gardens lie on the slope beneath the castle, an ideal place for artists. Three of these terraced gardens, built for the nobility after Italian models, can be visited. The entrance is next to the **Palais Kolovrat** (no. 10).

The Valdštejnská ul. and Waldstein Square *(Valdštejnské nám.)*, into which it leads, border the broad complex of the **Palais Waldstein** on two sides. This first baroque palace in Prague was built between 1624 and 1630 for the famous (and infamous) general Albrecht von Wallenstein and was a worthy memorial to his ambition. This hero of Schiller's play *Wallenstein* had made his way to the top by skilful strategy and leadership on the one hand and by intrigues and treachery on the other. During the Thirty Years' War he enlisted under the Habsburg Ferdinand II.

He won many important victories for Ferdinand II. These brought the imperial generalissimo not only power but also, along with a ducal title, wealth, which as a court favourite he was particularly well placed to increase. This process was helped not least by his participation in a grandiose coin swindle, so that in the end he was able to raise his own private army. His rapid rise came to an equally rapid end when, in secret deals with the enemy, he initiated tactical manoeuvres which would have led him to the Bohemian crown. However, the emperor saw through him and had him murdered in Cheb in 1634.

The grandiose residence matches Wallenstein's political ambitions. It was intended to rival the Prague castle. He acquired the site for the building by buying up and dispossessing the inhabitants of more than 20 houses. Even the city gate had to go, in order to give the architects (all Italians) enough space to provide their patron with a palace featuring all possible luxuries available at the time. However, the rather restrained outer facade that faces the square does not give anywhere near the same impression as a visit to the palace gardens mentioned above.

The greatest pride of the householder – apart from an artificial grotto, an aviary and a pond – was the triple arched **loggia** *(sala terrena)*, richly decorated with frescos. Today it serves as the podium for the open-air concerts held here in the summertime. The bronze statues of mythological gods and goddesses, scattered about the garden, are the work of Adriaen de Vries, court sculptor to the Emperor Rudolf II. They are, however, copies – the originals were taken to Sweden as spoils of war in 1648 and are now located in the park of the Drottingholm palace near Stockholm. Another work by this Dutch sculptor is the figure of Hercules fighting the dragon in the middle of the small pond. The fountain with the sculpture of

At Eastertime, painted eggs are on offer everywhere in Prague.

Venus and Cupid is also remarkable. Opposite the loggia is the former **Riding School**, where art exhibitions are held nowadays.

The Tomášská ul. leads from Waldstein Square back to Malá Strana Square. Go past the house "The Golden Pretzel" (no. 12) and you will come to the baroque house "The Golden Stag". This house bears one of the most beautiful house signs in the whole of Prague. It shows St Hubert with a stag. The sculpture is the work of F. M. Brokoff.

Before house numbers were introduced, during the reign of Maria Theresa in the 18th century, these house signs were used to identify the houses. They were based on the profession or craft of the house owner, his status or the immediate environment of the house. Animal and other symbolic signs, both of a secular and a religious nature, were popular. If the owner changed, the house retained its original sign. Sometimes the new owner even took over the name of the house.

We recommend that you go back to Waldstein Square and from there, make a short excursion into the Sněmovní ul. This street and the adjoining cul-de-sac, *U zlaté studně* (The Golden Fountain) form a picturesque corner. Hidden away at the end of the little alley is a garden pub with the same name as the street. Also noteworthy is the Renaissance house **The Golden Swan** (*Sněmovní ul.* no. 10), which hides a beautiful inner courtyard. Go back in the direction of the Thunovská ul., which leads into the **Castle Steps**. These so-called New Castle Steps are not to be confused with the Old, which lead to the other end of the castle. The New are actually much older than the Old, but that's the way of things in Prague.

Neruda Alley: Parallel to the Castle Steps (Zámecké schody) lies the Neruda Alley *(Nerudova ul.)*, named after the famed Czech poet, author and journalist Jan Neruda (1834–91) who lived in the upper part of the street, in house no. 47, **The Two Suns**. His work is inspired by the everyday life of the Malá Strana. Incidentally, his name was adopted by the Chilean poet Ricardo Eliecer Neftalí Reyes y Besoalto, now the Poet Laureate Pablo Neruda.

Many of the middle-class houses in this street were originally built in a Renaissance style and later given baroque additions. They often bear house signs which don't match the names of the houses. For instance, house no. 6, **The Red Eagle**, has a sign showing two angels. In the case of house no. 12, **The Three Violins**, it is known that several generations of violin makers lived here. More signs can be seen on the houses **The Golden Chalice** (no. 16), **St John Nepomuk** (no. 18), and **The Donkey and the Cradle** (no. 25). A small pharmacological museum is housed in the former pharmacy, **The Golden Lion.**

As is so often the case in Prague, two

Left, all sorts of different views of Prague are for sale at the Charles Bridge. Right, typical steps in the Malá Strana.

embassies have settled into the two baroque palaces in this street. On the left is the **Palais Morzin** (Romanian Embassy). Its unusual facade ornament – the heraldic Moors which support the balcony, the allegorical figures of Day and Night and the sculptures representing the four corners of the world – are the work of F.M. Brokoff.

Somewhat higher up is the **Palais Thun-Hohenstein** (Italian Embassy), which is decorated with two eagles with outspread wings, and is the work of Matthias Braun. The two statues of Roman deities represent Jupiter and Juno. The palace is connected by two passages to the neighbouring **church and monastery of St Cajetan**, creating an architectural unity typical of the closing years of the 17th century.

From the end of the Neruda Alley you can get a splendid view of the **Palais Schwarzenberg**, mentioned above in connection with Hradčany Square. The alley gives way to a romantic stairway leading up to the castle; to the left, the Loretánská leads out of the Malá Strana in the direction of Strahov monastery. This is also the way to reach the Loreto Shrine (see page 164).

At the back of the last houses of the Neruda Alley a maze of courtyards lies hidden. They fall in a series of terraces into the valley between the two hills. At the bottom are a few alleys that have almost a village character. If you go back a little, you will reach the rococo **Palais Bretfeld** (no. 33), with a relief of St Nicholas on the portal. In earlier years famous balls took place in this building, which even Giacomo Casanova is said to have attended.

From here the steps Janský vršek lead down and then turn right into the Sporkova ul., which leads us along the slope mentioned above. It then curves and leads into the Vlašská ul., directly opposite the **Palais Lobkovic**. This magnificent baroque palace now contains yet another embassy: Germany's. The palace garden is partially open to the public and is worth visiting because of the view.

If you want to visit the parks of the Petřín Hill or the Strahov monastery, you have to follow the the Vlašská ul. to your right. However, you can also go on in the opposite direction and use the cable car, which runs between the Ujezd street and the peak of the Petřín Hill.

On the way lies the **Palais Schönborn** which now houses the US Embassy. Its splendid garden, which can be seen from the Castle Ramp, is not open to the public. However, you can visit the particularly lovely baroque terraced garden of the **Palais Vrtba** (*Karmelitská ul.* no. 25). It is a small garden, but full of atmosphere, and has a *sala terrena* and sculptures by Matthias Braun.

Further along the Karmelitská you come to the **Church of St Mary of the Victories** (*Kostel Panny Marie Vítězné*), the first baroque church to be built in Prague. It was built as a monument to the Counter-Reformation brought to Prague by the Habsburgs. The furnishings, all of a unified style, date from the 17th century; the saints' pictures by the altar are the work of P. Brandl. It is in this church that the **Infant Jesus of Prague** is kept. It is revered and believed to work miracles, and has achieved world-wide fame. It is a wax figure of Spanish origin and is always clothed in one or other of its 39 costly robes.

The Petřín Hill: It's not far from here to the **Petřín Cable Car** mentioned above. It also stops at the restaurant "Nebozízek", which has a magnificent view. It is known as a cable car, but the cars don't actually hang on cables, they run on rails. The fact that the contraption looks so curious is down to the original method of locomotion. The old cars had a water tank, which was always filled at the top and emptied at the bottom. In this way the cars going up were powered solely by the weight of the cars going down. The cable car was inaugurated in 1891. In the 1960s the water tanks were done away with and it is now powered by more modern means.

The park on the Petřín Hill was formed by linking up the gardens which had gradually replaced the former vineyards. On the level of the upper cable car station, a path offering a marvellous view leads all the way through the park to the Strahov monastery. Apart from the delightful view, the park also has other sights to offer.

Starting in the most southerly corner, visit the **Villa Kinsky** which houses the Folklore Museum. On the way, in a northerly direction, lies a little wooden church, a wonderful example of folk art of the 18th century. It comes from the Carpathian Ukraine and was rebuilt on this spot in 1929. The church was a gift from the inhabitants of the small village of Mukacevo in that region, which became part of Czechoslovakia after World War I and part of the Soviet Union after World War II. There is also a belfry from Wallachia standing next to the museum.

The **"Wall of Hunger"**, which you will discover further on, leads down the slope and is part of the fortifications which Charles IV had built. According to rumour, this project was undertaken to provide work for the starving and impoverished. Near the wall lies the **People's Observatory**, a popular meeting place of amateur astronomers from near and far.

On top of the hill is the **Observation Tower**, a scaled-down version of the Eiffel Tower, which is 197 feet (60 metres) high. It was built for the Prague Jubilee Exhibition in 1891. Not far from the tower is **St Laurence's Church** and a **labyrinth of mirrors**. The latter brings the tour to a pleasant end, particularly for children.

alais chönborn: owadays ome of the S mbassy.

The claim is often made that Bohemian cuisine is anything but healthy or easily digestible for Western stomachs. Once you have accepted the fact that vegetables and salads are in short supply, that in the plainer pubs and restaurants the relationship of meat to dumplings is about one to four, that paprika and pickled Prague gherkins are the standard accompaniment to any dish, and that boiled Prague ham is on offer almost everywhere in a hundred different varieties, you can safely go and explore the Prague restaurants. Only one more hurdle remains to be overcome before enjoying a hearty Bohemian meal – the waiter. It can happen that all the tables are "reserved", although no one is sitting anywhere, and there's no sign of anyone coming, either.

All the large hotels in Prague have several restaurants and snack bars. Each hotel in the top price bracket also has both a Bohemian and a French restaurant.

Eating Bohemian: However, where should you go to eat genuine Bohemian roast pork *(vepřová pečeně)* with Bohemian dumplings *(knedlíky)* and cabbage *(kyselé zelí)*? For quick service, value for money and a good meal go to **U Bonaparta**, Prague 1, Nerudova 29. With it, drink a beer (specific gravity 12 degrees) from Prague's own Smichov brewery. Also in the Bohemian tradition, only kosher, is the food served in the **Jewish Restaurant** in Prague 1, Maiselova 18. Nearby, at Maiselova 5, excellent food based on traditional recipes is prepared by the smallest of Prague's numerous wine bars, the **U Rudolfa II**.

If you want a more formal meal, with an aperitif and several courses, then you would do well to go to the **Pelikán**, Prague 1, Na příkopě 7. However, in restaurants such as this one, it is necessary to reserve a table. The decor is not over-lavish and the guests are not overloaded with enormous white napkins, but can relax and enjoy good food and Moravian wines. If your foreign language skills extend to German, but not Czech, the waiters will translate the menu to fluent German. In fact, most waiters in Prague can speak either German or English, often both.

Those wishing to combine their experience of Bohemian cuisine with an authentic vaulted cellar atmosphere can take themselves off to **U Sixtů** in the Celetná. Another recommended cellar restaurant is the **Halali Grill** on Wenceslas Square. This high-class establishment offers game specialities as well as a live gypsy band. Other places to go for game are the **Myslivna** in Prague 3, Jagellonská 21 and the venerable **Valdštejnská hospoda** (Waldstein Inn) at the foot of the Hradčany. If you would like fish for a change, we recommend the **Baltic Grill** in the arcade at Wenceslas Square 43.

The people of Prague like their food flambéed, and among the masters of this skill are the waiters in the former Ursuline convent in the Národní třída, the **Klášterní vinárna**. Here the delicious *palačinky*, thin pancakes filled with ice cream and fruit, are turned once more in blazing alcohol. It lies right next to the New Theatre and the National Theatre, and as such is a good address for a late supper after a visit to the opera. Surrounded by American, German and Italian diplomats, you can – how could it be otherwise among such a select clientele – eat an excellent meal in the **Lobkovická vinárna**, Prag 1, Malá Strana, Vlašska 17.

Left, an old Prague beer cellar in pre-1900 times. Above, breakfast tables in the Hotel Evropa.

International tastes: If you've had enough of dumplings and pork, and would prefer something a little more international, then the **Trattoria Viola** (Národní třída 7) offers a chance of escape. Here you will find an interesting combination of Italian cuisine and limited Bohemian ingredients. One further point in its favour: it serves Chianti wine. Indian cuisine can be enjoyed in the extremely formal top-class **Indická restaurace**, Prague 1, Nové město, Stěpánská 63. Fans of Chinese food should either head for the **Peking** (Legerova 64) or the ever popular **Cínská restaurace** in the Nové město, (Vodičkova 19), though it is not unusual to have to book three or four days in advance here.

Fast food: The recently opened **Arbet** now caters for fast food fans. Further details are unnecessary, all you have to do is to follow the trail of red and gold chip tubs that litter the Na příkopě. And if you need to fill a corner, you can always buy a fried sausage from the stalls in Wenceslas Square, which are open until late at night.

STARÉ MĚSTO – THE OLD TOWN

The Old Town (Staré město) of Prague is spread along the right bank of the Vltava and around the Old Town Square. The main streets – Národní třída, Na příkopě and Revoluční – mostly follow the course of the city fortifications, which no longer exist. The name *"Na příkopě"*, which means "by the moat", indicates that they were built on the site of the old moat which separated the Old Town from the New. Together these two districts form the actual city centre of Prague.

However, the Old Town has kept its individual character. The pattern of streets and squares has remained largely unaltered since the Middle Ages. Originally the Old Town lay some 6 to 9 ft (2–3 metres) below the modern street level. The area was, however, subject to repeated flooding, which is why the street level has been raised bit by bit since the late 13th century. Many houses have Romanesque rooms hidden in their basements.

The historic core of the Old Town is built on these foundations, and every age has left its signs for us to read, whereby the overwhelming influence of the baroque cannot be overlooked. However, it only finds its expression in individual buildings and has not altered the basic structure of the district. The only large intrusion is the massive building of the Jesuit College, the Clementinum. Here and there you can also see traces of the 19th and 20th centuries, for the development of the river bank gave the big city a chance to break in. However, apart from demolition of the Jewish Quarter, the entire district has hardly lost any of its charm.

The present-day appearance of the streets is still marked by the unceasing succession of houses with the most varied facades. This is a living district with a well-balanced mixture of homes, offices, shops, small businesses, several schools and leisure facilities, all of which play a part in forming the impression given by the district.

The first settlements on the site of the Old Town for which there is any historical evidence date from the 10th century. They concentrated around the crossing of three important trade routes, which met at the ford across the Vltava, a little downstream from where the Charles Bridge stands today. According to a contemporary report, a large market place with numerous stone houses covered the site of the present Old Town Square. As the years went by this market place grew, and was fortified with a city wall in the early 13th century. Around 1230, the settlement received its city charter. By this time it was possible to speak of a large town, in European terms. In 1338, John of Luxembourg granted the citizens of the Old

Preceding pages: Pilsner, Budweiser. Below, the Jan Hus memorial.

Town the right to their own Town Hall, and in the years that followed, during the rule of Charles IV, the city experienced an immense economic and cultural boom.

The **Charles University**, the oldest university in Central Europe, was founded in 1348. Even if the importance of the imperial residence did diminish later on, the Old Town kept its leading position in Prague. When the four independent cities became one administrative unit in 1748, it was the Town Hall in the Old Town that became the seat of the administration.

Old Town Square: All the busy streets near the border of the New Town lead the visitor who crosses over into the Old Town to the Old Town Square (Staroměstské náměstí). The streets lead to the square from all sides like the rays of the sun and make it a natural centre. The **memorial** in the middle of the square honours the great reformer Jan Hus and was erected on the 500th anniversary of his death (6 July 1915). In more recent years the Old Town Square has also been the scene of large gatherings.

The houses on the east side of the square form a singular backdrop. This juxtaposition of the most varied building styles is typical for the Old Town and, together with the towers of the Týn church, it gives the Old Town Square its special character. To the left you can see the **Palais Kinsky** with its late baroque facade, which already incorporates some rococo elements. It was built by A. Lurago, according to the plans of Kilian Dientzenhofer. Today you will find the National Gallery's collection of prints and drawings here.

To the right of the palais is the Gothic house **The Bell** *(Dům u Kamenného zvonu)*, which has recently been restored and had its original facade replaced. Various exhibitions are held in the interior, which is worth seeing. The

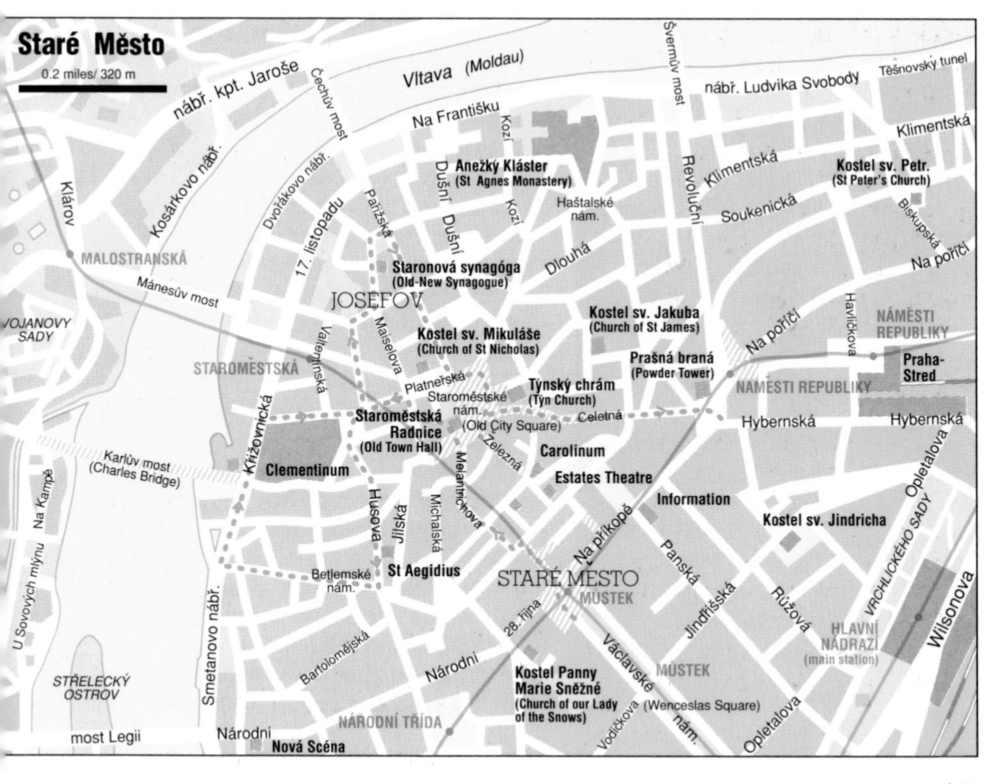

two neighbouring houses are connected by an arcaded passage with ribbed vaulting. To the left, the **Týn School**, originally a Gothic building, was rebuilt in the style of the Venetian Renaissance. Attached on the right is the early neoclassical house **The White Unicorn**.

You can gain access to the **Týn Church** through the Týn school or from Celetná ul. no. 5. It was the third church built (in 1365) on this site, the successor to a Romanesque and an Early Gothic building. Up until 1621 it was the main church of the Hussites. The tall nave received baroque vaulting after a fire. The paintings on the high altar and on the side altars are by K. Škréta, the founder of Bohemian baroque painting. Other remarkable works of art are the Gothic Madonna (north aisle), the Gothic pulpit and the oldest remaining font in Prague (1414). To the right of the high altar is the tombstone of the Danish astronomer Tycho Brahe (1546–1601) who worked at the court of Rudolf II. The window immediately to the right of the south portal is a curiosity – through it you can see into the church from the neighbouring house. One of those who could by this means experience the services without leaving his apartment was, for a time, the author Franz Kafka.

The baroque **Church of St Nicholas** (Kostel sv. Mikuláše) on the other side of the square is also the work of Kilian Dientzenhofer. The statues on the facade are by A. Braun, a nephew of M.B. Braun. The unusual proportions of the church have come about because houses originally stood in front of the building, completely separating it from the square. It is interesting to see how the architect has succeeded in creating so perfect a building in such a relatively small space. The house to the left of the church, by the way, is the birthplace of Franz Kafka.

Old Town Hall: In times gone by, the area of the small park opposite the church was occupied by a Neo-gothic wing of the **Old Town Hall** *(Staroměstská radnice)*. It was destroyed in the last days of World War II. If you walk around the Town Hall Tower, which protrudes into the Old Town Square, you gain an unobstructed view of the historic part of the Town Hall. At first, the house next to the tower on the left was purchased by the citizens of the Old Town and declared a town hall. Later, three further houses in this row were acquired.

The tower was built in 1364 and later had the oriel chapel added. The **astronomical clock** on the tower dates in its earliest form from 1410. It consists of three parts. In the middle is the actual clock, which also shows the movement of the sun and moon through the Zodiac. The representation is in accordance with the geocentric views of the time. Underneath is the calendar, with signs

Left, the Powder Tower. Right, the famous Clock on the Old Town Hall.

4
5
6
7
8
9
10
11
12
OCCASVS

of the Zodiac and scenes from country life, symbolising the 12 months. The artistic work on the calendar is by the well-known Czech painter Josef Mánes.

The upper part is a popular attraction. Every day on the hour the figures play the same scene: Death rings the death knell and turns an hourglass upside down. The 12 apostles proceed along the little windows, and a cockerel flaps its wings and crows. The hour strikes. To the right of Death a Turk wags his head. The two figures on the left are allegories of greed and vanity.

Also on the hour a guided tour leaves for a trip through the historic rooms of the Town Hall, which also contains exhibition rooms.

The memorial tablets on the Town Hall Tower are reminders of some of the important events that have taken place in this square: the execution of the radical Hussite preacher Jan Želivský (1422), who became known to history through the first Defenestration of Prague; the executions of the "27 Bohemian gentlemen" (1621), a punishment of the leaders of the rebellion of 1618, intended to serve as an example to others; the liberation of Prague by the Red Army on 9 May 1945.

Side streets: Passing the house **U Minuty**, you come to the **Little Square** *(Malé náměstí)*, evocative of medieval Prague. Surrounding the fountain with its pretty Renaissance railings are a number of fine houses, each with its own history. In 1353 in House No. 11, a Signore Agostino of Florence established the first documented apothecary in the city, and during the reign of Emperor Charles IV House No. 1 was the home of a herbalist from Florence. Most spectacular is the **Rott House** (No. 3), whose cellar was once the lower floor of a Romanesque town house. The first Czech bible was printed here in 1488; at

In summer people in Prague like to sit outdoors.

the turn of the century the new owner, an ironmonger, had the building renovated, painting the facade with the original sign of three white roses and a selection of his wares.

In order to enjoy this change of atmosphere, you should make a short trip into the neighbouring little streets – the Karlova ul, for instance, which bends to the left, and then straight on into the Jilská ul. Soon you will see on your left house no. 18, which in earlier years bore the name **Two Stags with one Head**. An unassuming arch is followed by a passage to the Michalská ul. (the so-called "Iron Gate"). It links up with another passage through a palace courtyard with Renaissance arcades. However, there is also another route on the left, which crosses the courtyard of a monastery containing **St Michael's Church**, in which Jan Hus preached. Both ways meet up again in the Melantrichova ul. just before the Kožná ul. leads into it. The first house on the left, **The Two Golden Bears**, is a beautiful example of Renaissance architecture. The Kožná ul. leads you out of the labyrinth back to the Old Town Square.

Celetná ul. and Powder Tower: The Celetná ul. is named after the medieval bakers (calty) of small loaves. It is one of the oldest streets in the whole of Prague, for its course follows the line of the old trade route to the east where it left the Old Town markets.

Following the large-scale restoration undertaken in recent years, most of the baroque facades in this model street shine in new glory. Of particular architectural interest is the Late baroque **Palais Hrzán** (no. 12). Nearby is the wine bar **The Golden Stag** (*U zlatého jelena*), which is situated in what were originally the rooms of one of the oldest stone houses in Prague. An architectural rarity of quite a different order is the unique Cubist house **The Black**

edding oach in ont of the ld Town all.

Madonna (no. 34), designed by Joseph Gocar (1911–12).

At the end of the Celetná ul. is the Late Gothic **Powder Tower** (Prašná Brána). It was built in the second half of the 15th century as an impressive city gate, replacing an older gate which had previously stood on this site. Its special status among the 13 gates of the Old Town fortifications came about because the Royal Court (no longer in existence), which acted as the royal residence in the 15th century, was right next door. It acquired its name in later years, when it was used as a gunpowder magazine. The Neo-gothic roof was added during rebuilding in the second half of the 19th century. The tower is open to the public on Wednesdays and weekends from April to October and offers an interesting view.

On the site of the Royal Court mentioned above, the **Municipal House** *(Obecní dům)* was built in the years 1906–11. The splendid Art Nouveau building was created in response to the politically and economically strengthened national consciousness of the Cezch bourgeoisie around the turn of the century. A whole generation of Czech artists worked on this building. Appropriately enough, it was here that the Czech republic was declared in 1918, after World War I. The interior, which is still in its original condition, can be seen in the restaurant and the café belonging to the building. There are also various rooms which are used for social events and the **Smetana Hall**, a famous concert hall.

Another example of Prague late Art Nouveau (similar to the Viennese Secessionist style) is the **Hotel Paříž**, which is situated next door to the Municipal House.

If you enter the little alleys at the back of these buildings, you soon come to the **Church of St James** *(Kostel sv Jakuba)*

The Towers of the Týn church.

in the Malá Štupartská ulice. You can also get to it from the Celetná ul., through one of two passages in the houses no. 17 and no. 25. Like so many churches in Prague, St James's, which was originally founded by the Minorites during the reign of Charles IV, was rebuilt several times until it attained its present baroque form. Notable works of art are the reliefs on the main portal, the ceiling frescos and the painting by V. V. Reiner on the high altar. Particularly valuable from an artistic point of view is the tomb of Count Vratislav Mitrovic, the work of J. B. Fischer von Erlach and F.M. Brokoff. The almost theatrical quality of the interior provides a fine stage for the frequent organ concerts held on the ornamental and powerful instrument dating from 1705.

The cloisters of the former Minorite monastery adjoin the north side of the church. Musical instruments of all kinds can be heard in the former monks' cells in the upper storey, for the monastery is now a music school.

he best ew of the elantrich- va is from e Old own Hall.

Between St James's and the Týn church lies the Týn Court, also known simply as the *Týn*. It is a peaceful place, separated from the rest of the city, with plenty of atmosphere. Originally it offered protection to foreign merchants. The whole complex, whose origins go back to the 11th century, is currently being renovated. You can use the Týnská ul. to get back to the Old Town Square. It leads around the Týn Court to the right to the north portal of the Týn church. The covered end of the street, the Týn Court gateway and the church portal with the magnificent tympanum from Peter Parler's workshop together make up one of the most picturesque corners of the Old Town.

Pařížská to Crusader Square: The impressively proportioned Pařížská třída (Paris Street) leaves the Old Town Square by the church of St Nicholas and leads to the sights of the former Jewish Quarter (Josefov). You should combine a tour of this Jewish district with a visit to the **St Agnes Convent** *(Anežsky klášter)*, which lies a little out of the way in the Anezská ul. The convent is the first Early Gothic building in Prague (founded 1234). The whole complex, which included two convents and several churches, fell into decay over the years and parts of it were completely destroyed. After painstaking work lasting many years, restorers succeeded in bringing some rooms back to their original state. These were linked up to form the present-day historic complex by means of carefully reconstructed additions. The convent contains an exhibition of the Craft Museum (19th-century crafts) and a collection from the National Museum (Czech painting of the 19th century).

Leaving Josefov via the street Ul. 17 Listopadu, you will have only a few steps to go before coming to the **Arts**

and Crafts Museum (*Umělecko-prumyslové muzeum*), which has a public reference library and an exhibition room displaying a variety of objects from various crafts and skills.

Located diagonally opposite the museum is the **Rudolfinum** (also known as *Dům umělců* or House of Artists), an impressive building in Neo-renaissance style, which faces the square, *nám. Jana Palacha*. Nowadays it is the seat of the Czech Philharmonic Orchestra and boasts a magnificent concert hall. From 1919 to 1939 it served as a parliament building.

We recommend that you cross the square now and enjoy the view of the Charles Bridge, the Malá Strana and the castle from the banks of the Vltava. If you carry on upstream along the embankment and then follow the Křižovnická ul. you will pass the massive and somewhat gloomy facade of the Clementinum and come to the **Crusader Knights Square** (*Křižovnické nám.*) with its monument to Charles IV.

Right on the first pillar of the **Charles Bridge** is the **Old Town Bridge Tower**, built, like the bridge itself, by Peter Parler. The remarkable statues which ornament the tower are also from the great master's workshop.

The baroque **Church of the Crusader Knights** *(Kostel sv. Františka Serafinského)* on the river bank side of the square is dedicated to St Franciscus Seraphicus. In the earlier years, the church belonged to the monastery of the "Order of Crusader Knights of the Red Star", the sole Bohemian knightly order at the time of the Crusades. The magnificent cupola of the church is decorated with the fresco *The Last Judgment* by V. V. Reiner.

The houses built into the river were originally the Old Town water mills. The last house, which is decorated with sgraffito and is a former waterworks, now contains the **Smetana Museum**.

Opposite the Bridge Tower you can see the baroque facade of **St Saviour's Church** *(Kostel sv. Salvátora)*, which is part of the Jesuit college of the **Clementinum**. This broad complex was founded by the Jesuits, who were called to Prague in 1556 to assist the return of the country to the Catholic fold. Nowadays the State Library of the ČSFR is based here.

Charles Alley: The narrow and twisting **Charles Alley** *(Karlova ul.)* has always been the link between the Charles Bridge and the Old Town Square. In house no. 4 the astronomer Johannes Keppler lived for a while. A little further on, in house no. 18 **The Golden Serpent**, the Armenian Gorgos Hatalah Damashki opened the first Prague café in 1714.

Leave the Karlova, following the outer wall of the Clementinum, and you have only a few steps to go to the Nám. primátora dr. V. Vacka. The Art Nouveau building in this square is the

Left, friendly young man in uniform. Right, everyone is waiting for the cock to crow.

New Town Hall with the residence of the Mayor of Prague. The **Clam-Gallas Palais** lies at the corner of the Husova ul. It is a magnificent baroque building, designed by the Viennese court architect J.B. Fischer von Erlach. The portal ornamentation is by Matthias B. Braun. Opposite the Palais is the **City Library**, which also houses a department of the National Museum (Czech art of the 20th century).

To the south of the Karlova ul. a network of tiny streets spreads out. They invite you to stroll along them.

Follow the Husova ul. for now. On the right you will notice a facade built in the style of the Venetian Renaissance. This house (no. 19) contains the **Central Bohemian Gallery**, which holds exhibitions of regional art. A little further to the left is the **Church of St Aegidius** *(Kostel sv. Jiljí)*. You get the best impression of the clear, Gothic lines of exterior of the church, which contrasts with the overloaded baroque interior, if you simply walk around it. The central courtyard of the monastery that adjoins the church to the left has an atmosphere of quiet solitude.

A road runs off at right angles to your right, and leads to the Retě zová ul. House no. 3, *Dům pánů z Kunstátu*, should be mentioned. In its basement the whole ground floor of a Romanesque palace has been preserved. These rooms are open to the public and serve as exhibition rooms for the "Prague Centre for the Preservation of Monuments". Further on in the same direction is the little square *Anenské nám.*, which has a cosy atmosphere.

An important memorial to the Hussite past is the **Bethlehem Chapel** *(Betlémská kaple)* in the square of the same name *(Betlémské nám.)*. Today's building is a faithful reconstruction of the original chapel founded in 1391. Here the mass was said in Czech. The

Christmas tree in the Old Town Square.

plain interior had the pulpit as its focal point and not the altar. It could hold up to 3,000 people. In the early 15th century, the famous reformer Jan Hus preached and worked here. His ideas spread out from this place to all over the country. In 1521, the leader of the German peasants' revolt, Thomas Münzer, also preached in this church.

A picturesque courtyard located right on the western side of the square contains the **Ethnological Museum** *(Náprstkovo muzeum)* with an exhibition of artifacts from Asian, African and American cultures.

Just a little out of the way, on the corner of Ul. Karolíny Světlé and Konviktská ul., lies the **Holy Cross Rotunda** *(Rotunda sv. Křížě)*, a Romanesque round church dating from the beginning of the 12th century.

A curious church building awaits you in the Martinská ul. Originally Romanesque, later rebuilt in Gothic style, the **Church of St Martin in the Walls** was incorporated into the city walls. Here, in 1414, Holy Communion was first administered "in both forms", (i.e. both bread and wine given to the laity).

Our tour of the Old Town ends back in the neighbourhood of the Old Town Square. The Martinská ul. leads into the **Coal Market** *(Uhelný trh)*, and some old market streets follow. Worth further inspection is the picturesque alley *V kotcích*, in which time seems to have stood still. Between the **Fruit Market** *(Ovocný trh)* and the Železná ul. lies the **Estates Theatre**. It was opened in 1783 as the Nostitz Theatre and is the oldest theatre building in Prague. The theatre played a large part in the cultural life of the city. To the left of it lies the **Carolinum**, a historic building which belongs to the Charles University. The magnificent oriel in the outer wall is a remnant of the original 14th-century Gothic building.

very ohemian a usician".

Der Anfang des dreiſsigjährigen Krieges.

10 March, 1948, early morning. The caretaker in charge of heating the Czernin Palace, Karel Maxbauer, found his employer, the foreign minister of the Czech republic, Jan Masaryk (aged 63), dead in the courtyard. Masaryk, son of Tomáš Masaryk, the founder of the Czech Republic, was the only non-communist cabinet member left after the "bloodless" coup in February 1948. One month before, 12 non-communist ministers in the coalition cabinet of the communist premier Clement Gottwald had resigned. Under pressure from communist militia and the threat that Soviet troops might invade, state president Beneš appointed a communist cabinet. The sole exception: Jan Masaryk. Was the fall from the bathroom window, 45 feet (15 metres) up, suicide?

23 May 1618. Enraged Protestant citizens of Prague threw three Catholic councillors out of the window of the Hradčany castle into the moat. This confrontation between Catholics and Protestants led to the Thirty Years' War. In order to consolidate his power in Bohemia, Ferdinand II had originally agreed to the terms of Rudolf II's "Letter of Majesty" which guaranteed religious freedom and the unrestricted building of churches. But Ferdinand II soon showed himself to be a supporter of the Counter-Reformation. His harsh and merciless actions led to open revolt. Protestant churches in Bohemia were closed, and some were even torn down.

The "Defenestration" by the enraged Estates of Prague had far-reaching consequences. The monks of the Strahov monastery and archbishop were exiled, Ferdinand II was declared deposed in 1619 by the Bohemian Estates. In 1620, the army of the Estates, dependent on foreign help and led by Frederick V of the Palatinate, was defeated by the imperial army in the Battle of the White Mountain near Prague. The power of the Habsburgs and the Catholic church was re-established. In the subsequent "Bloody Trial" in Prague, 22 Czech and five German noblemen were publicly tortured and executed in the Old Town Square.

July 30, 1419. A stone was thrown from the window of the New Town Hall at a procession of armed Hussites. The unavoidable, long-smouldering conflict finally erupted. The Hussites stormed the Town Hall and threw three consuls and seven citizens out of the windows. The Hussite Wars had begun.

Jan Hus was born in 1369 in Husinec, Bohemia. He became a priest in 1400, in 1402 he became Rector of the Bethlehem Chapel, and in 1409, Hus was made Rector of Charles University. His fiery speeches attacked the worldliness of the Catholic church, the immoral behaviour and lack of sanctity of the clerics, and promoted a Czech national movement.

Hus's ideas were popular with many of the people and with King Wenceslas IV. He was forbidden to preach by the Archbishop of Prague, and excommunicated in 1411 when he refused to obey. Hus agreed to attend the Council of Constance if granted a safe conduct, and obtained one from Wenceslas's successor Sigismund. However, he was arrested and imprisoned, charged with heresy and burned at the stake in 1415.

Left, portrayal of the Defenestration of Prague, 23 May 1618. Above, Jan Masaryk is dead.

DIVADLO
POLLENA
polská kosmetika
BULGAR

NOVÉ MĚSTO – THE NEW TOWN

Nové město is the New Town district of Prague, and has not nearly as many interesting sights to offer as the other districts described above. However, you should not miss taking a stroll through this district, even if you do have to cover considerable distances on foot. This is one district where you will get a good impression of everyday life in Prague. The shops do not as yet cater for tourists, and the condition of most of the houses proves that more than the Old Town Square is in need of renovation.

Around Wenceslas Square: The gently rising, gigantic former Horse Market is crowned by the martial-looking equestrian **Statue of St Wenceslas**, finally erected by Josef Myslbek in 1912 after taking 30 years to plan and design. The people of Prague congregate when necessary in Wenceslas's shadow, as well as in that of Jan Hus in the Old Town. This is where proclamations are made and demonstrations are started. Huge crowds assembled here in 1919 and in 1939. During the Prague Spring of 1968, and again during the "velvet revolution" of 1989, pictures of the Square were transmitted all around the world.

With a length of almost half a mile, the scale of Wenceslas Square *(Václavské nám.)* is overwhelming. Nowadays, the square is dominated by its hotels, pubs, restaurants, cafés, banks and department stores. It is a gigantic and busy boulevard, along which half the inhabitants of Prague seem to stroll in their leisure time, joined by masses of tourists. What you can't find in Wenceslas Square, it's said, won't be found anywhere else in the Czech Republic.

The old two-storey baroque houses that once lined the square have gone. They have been replaced by six and seven-storey buildings, of which only a few, such as the **Hotel Evropa**, still retain their Art Nouveau facades. Behind the statue of St Wenceslas, so redolent of history, the square is enclosed by the **National Museum** *(Národní muzeum)*. A successor to the National Theatre in the Národní třída, it was built in 1885–90 by the Prague architect Josef Schulz. Although he was assistant to Zitek, the architect who designed the National Theatre, Schulz wasn't quite a match for Zitek with this copy. The building that was intended to become the spiritual and intellectual centre of the Czech nation seems rather unfortunate and clumsy.

Next door, replacing the old Produce Exchange, is a conspicuous glass building, the **New Parliament Building.** Almost in the shadow of the big Parliament Building lies Wilsonova 8, the **Smetana Theatre** *(Smetanovo divadlo)*. It was built in 1888, a successor to the wooden "New Town Theatre" which had stood on the same site. At

Preceding pages: view from the National Museum across Wenceslas Square. Below, equestrian statue of St Wenceslas.

that particular time it was the "New German Theatre", the second largest German language stage in Prague and for that reason, it was not entirely without its problems.

If you walk down Wenceslas Square, the main shopping area of Prague begins about on the level of the streets Jindríšská and Vodičkova. Most of the expensive shops, department stores and bookshops can be found around the Metro station *Můstek*, in the pedestrian precinct Na příkopě and the 28. října.

At the lower end of the square two streets (both for pedestrians) branch off. The name of the Metro station *Můstek* is a reminder that a bridge which led to the Old Town once stood on this site. Remains of the bridge can be seen in the underground station. The pedestrian precinct of Na příkopě ("by the moat") follows the course of the old fortifications towards the Powder Tower. Don't waste too much time on the less than interesting displays in the shops. More interesting are no. 12, the **Palais Sylva-Tarouca** (built in 1670 and extensively altered in 1748), and no. 22, which dates from the 18th century and is now the **Slavonic House**, formerly the Palais Přichovský, then the German House.

Right opposite is the **Municipal House** *(Obecní dům)*, built in a Seccessionist style. The passage from Na příkopė 11, next to the Café Savarin, and to Wenceslas Square is also attractive. Panská ul., in which the accommodation office of **Čedok** is situated, leads off from the Na příkopě.

On the other side, the street 28. října leads to the Náměstí Jungmannovo with its memorial to the Czech linguist Josef Jungmann (1773–1847). Here, many visitors to Prague go straight past the gate of the Franciscan rectory and overlook the church of **St Mary of the Snows** *(Kostel Panny Marie Sněžné)*, which was planned to be a massive

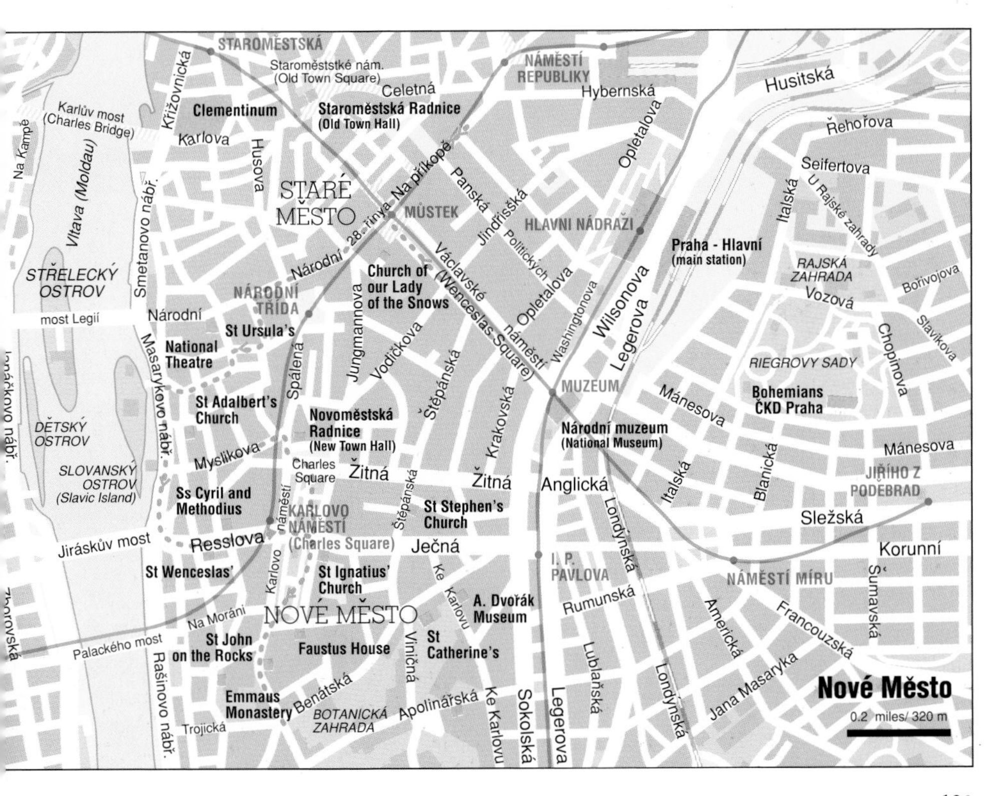

building. Today all you can see of the church, which was founded as a coronation church by Charles IV in 1347, is the choir. The plans envisaged a church comparable to St Vitus' Cathedral, a three-aisled Gothic cathedral church, and to become the tallest building in Prague. However, shortage of money and the start of the Hussite wars saw to it that the plans were never fulfilled. This is why the proportions of the church look rather odd. Inside, the 16th-century altar and the font dating from 1459 are worth special attention. You can get a good view of the church from the little park behind the Dům sportu, which leads to the Alfa Arcade and so on to Wenceslas Square.

The Národní třída leads off the Jungmannova nám. and on to the Vltava and the **Bridge of Legions** (Most Legií). Notable are the glass facade of the **New Theatre** *(Nová scéna)* and the **National Theatre** *(Národní divadlo).* Before you get to the National Theatre, and just before the New Theatre, you will pass the **Ursuline Convent** and the **Ursuline Church**. The church has just been restored, and in front of it is a group of statues, and notably among them, is St John of Nepomuk with cherubs, by Ignaz Platzer and dating from 1746–47. Nowadays the former convent buildings contain an excellent wine bar.

The National Theatre: The *Národní divadlo* is, purely and simply, the symbol of the Czech nation. In 1845 the Estates, with their German majority, turned down the request for permission for a Czech theatre. Money was collected, and the building of a Czech theatre declared a national duty. In 1852 the site was bought, and the foundation stone was laid in 1868. The building, designed by Josef Zitek in a style reminiscent of the Italian Renaissance, was opened in 1881 with a per-

The National Theatre is mirrored in the glass of the New Theatre.

formance of Smetana's *Libuše*. In August 1881, only two months after the opening, the National Theatre was burned down. Under Josef Schulz's direction, it was rebuilt and re-opened in 1883.

Right opposite the National Theatre, in the former **Palais Lažanský**, is the **Slavia**, one of the last great coffee houses of Prague.

Enroute to Charles Square: Take a look from the Bridge of Legions, which as the second oldest bridge in Prague has often had to change its name, up the Vltava, and you will see the Slavic island *(Slovanský ostrov)* with the Sophia Hall and the **Café Mánes**. At its upper end, near the Jiráskek Bridge, is the end of the Resslova, in which lies the Orthodox church of **SS Cyril and Methodius** *(Kostel sv. Cyrila a Metoděje)*. The starting point for boat tours on the Vltava lies between the **Jiráskek** and **Palacký** Bridges. You can get into the (usually closed) church through the former sacristy. After the assassination of the "Reich Protector" Reinhard "the hangman" Heydrich on 17 May 1942, three of the assassins and four other members of the Resistance barricaded themselves into the crypt. Their hiding place was discovered on the 18 June and they were removed by the SS after a bitter struggle.

Today a number of photographs and documents are displayed in the crypt and it serves as a memorial. The Nazis exacted cruel revenge for the assassination: Heydrich's successor Karl Hermann Frank ordered the village of Lidice near Prague to be burned to the ground. This was done on 10 June 1942 and all 199 men in the village were shot. The women were deported to concentration camps and the children were dispersed around Germany to be raised as Germans. Many of the women ultimately died in the gas chambers.

ɛft, the ational ıeatre and e New ıeatre. ight, in the AJ epartment ore.

If you turn off from the Resslova into the Na Zderaze and then into the Na zbořenci, you will come straight to the Kremencova, which boasts one of the most famous of Prague's beer pubs, the **U Fleků**, a brewery and small restaurant which serves its own special dark beer – something not to missed. U Fleků, together with U Kalicha (The Chalice), is probably the pub most often visited by tourists in the peak season in Prague. Every room has a different name. For instance, you not only find the **Velký sál** (Great Hall), but also the Jitrnice, which means "liver sausage". The shady beer garden is extremely popular in summer, and in the evenings a small brass band adds to the atmosphere. A traditional Czech cabaret appears every evening in the U Fleků and is one of the last of the Prague cabarets.

Charles Square: Your route through the New Town district of Prague continues from the Resslova to the Karlovo náměstí, or **Charles Square**. This, too, was part of Charles IV's building project for the New Town and was laid out in 1348. Charles Square was the biggest market in the city and was, until 1848, known as the "Cattle Market". Its present appearance is due to 19th century rebuilding. The monuments in the park portray famous Czech scientists, scholars and literary figures.

More interesting than the square itself is the **Town Hall** situated *(Novoměstská radnice)* at the northern end. It was built in several stages between 1348 and 1418, after the founding of the New Town. Alterations to the south wing followed a hundred years later, in 1520, and the tower was rebuilt in 1722. The extensive renovations carried out in 1906 restored the building to its original splendour. This is where the first Defenestration of Prague took place, initiating the Hussite wars which lasted for 15 years.

A cause of some controversy in Prague – the MAJ department store.

In the middle of Charles Square, on the eastern side, is the **Church of St Ignatius** *(Kostel sv. Ignáce)*. This is where the Jesuits built their second college in 1659, following the Clementinum. The college was dissolved in 1770 and since then the large complex has served as a hospital. The church, built in 1665–70 by Carlo Lurago, was extended in 1679–99, with a pillared hall and arcade designed by Paul Ignaz Bayer. Inside, among other works of art, is an altarpiece *Christ in Prison* by Karel Škréta (1610–74).

A walk through the New Town doesn't offer nearly as much variety as a tour of Malá Strana or the Old Town. If you want to see interesting buildings and sights here, you'll need good, durable shoes. The facades of the New Town apartment houses are not really all that exciting. The soot that has been falling for decades onto the city is thick on the walls, and so everything looks a little dilapidated. Large-scale renovations, like those in Charles Alley, for instance, are not in evidence here. Even the present-day pharmacy which achieved fame as the **Faustus House** *(Faustův dům)*, is a sorry sight. In this building, originally a Renaissance house, at the bottom end of Charles Square, the alchemists Edward Kelly and Ferdinand Antonín Mladota conducted their experiments. The latter also entertained his guests with a series of conjuring tricks and a magic lantern.

culpture in ont of the etro ation árodní ída.

In Prague, a city sensitive to such activities, that provided reason enough to give the house its peculiar name. Kelly, no more a serious scholar than Mladota, was supposed to discover the Philosopher's Stone for Rudolf II, but did not succeed. Rudolf had the Englishman, whose ears had been cut off in his own country as a punishment for fraud, thrown into a cell. The search for the Stone probably lasted too long for Rudolf's liking. Kelly died of poison in prison, after two attempts at escape.

There are fewer legends surrounding the **Church of St John on the Rock** *(Kostel sv. Jana Na skalce)* in the Vyšehradská, just around the corner. It is a pity that it is so difficult to get into this very beautiful baroque church. Like many churches in Prague, it is locked for most of the time. A similar such case is the former monastery opposite, the **Emmaus Monastery** *(Kláster Na Slovanech)*. This establishment goes back to a foundation of Charles IV in 1347 and was destroyed by bombs in 1945. To replace the two towers, two sail-shaped buttresses were added to the church in 1967. They can be counted among the rather few examples of originality in modern architecture in Prague.

Unfortunately the whole structure is an illusion. The concrete additions of František Černý cover an unhappy ruin. In the cloisters right next door are Gothic frescos dating from the 14th century. The Emmaus monastery became famous as a medieval scriptorium producing Slavonic manuscripts. Today it is the home of a scientific institute, and the frescos can only be admired during office hours.

Only a few steps away from the Emmaus monastery lie the **Botanical Gardens** with their beautiful greenhouse, which is unfortunately also in need of renovation.

Around New Town: Keep following the Vyšehradská, which, once past the Botanical Gardens, turns into the Na slupi, and you will come to another important (but most often locked) church, *Maria Na slupi*. This former convent church of the nuns of the Elizabethan order is also Kilian Ignaz Dientzenhofer's work, and a rare example of a Gothic church supported by a central pillar.

If you go up the steps of the Albertov, you will come to another building that's well worth seeing, **Charles Church** *(Kostel sv. Karla Velikého)*. The former Augustinian monastery is surrounded by University buildings, and you enter it through a plain gate. It is evident just by looking at the exterior that the Charles Church is an unusual building. The edifice has an octagonal ground plan and a central dome. It was founded in 1350 by Charles IV and dedicated to Charlemagne in 1377, and is reminiscent of the imposing Imperial Chapel in Aachen.

The Charles Church lies right on the edge of the descent into the Nusle valley, and only a few yards away from its surrounding wall the **Nusle Bridge** arches across the valley, over the apartment houses of Nusle that lie beneath it. Its 1,640 feet (500 metres) make it the second longest bridge in Prague. There are two modern glass "palaces" at the end of the bridge. One is the new **Hotel Forum**, completed in 1988, and on the other side is the **Palace of Culture** *(Palác kultury)*, one of the "most nota-

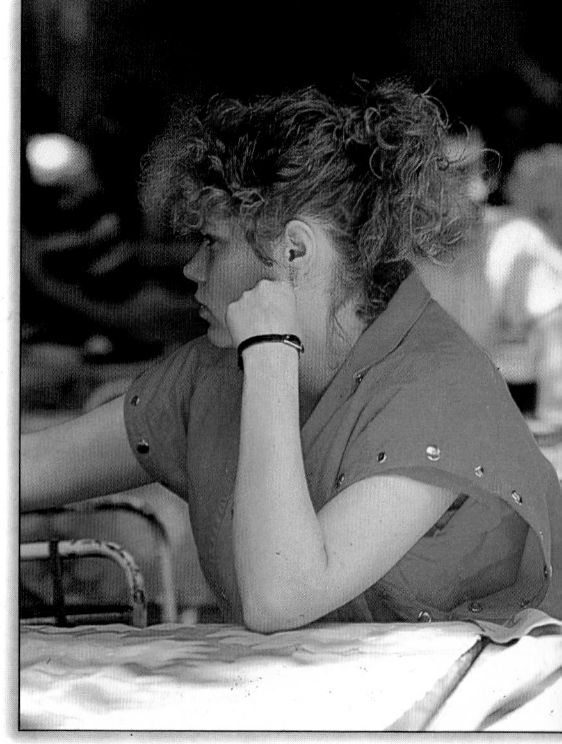

Left, a short spell of sunbathing during lunch break. **Right**, scene in the U Fleků.

ble of modern buildings", officially opened on 2 April 1981.

If you walk slowly back down the Ke Karlovu and past the Charles Church, you will see on the right-hand side the **Villa Amerika**, the Antonín Dvořák museum. This small, charming building was constructed in 1717–20 as a summer palace for the Michna family. Amidst the rather dreary university buildings the little baroque villa, named after a former 19th-century pub, provides a pleasant change for the eyes.

The **Lapidarium** opposite, in the **Church of St Catherine**, which can be reached by following the Kateřinská, is, however, less attractive. The church also goes back to a foundation made by Charles IV in 1355. It was largely destroyed in the Hussite wars. The octagonal tower is all that remains of the building from that time. The present-day building is the result of rebuilding work undertaken in 1737. The collection of cast figures and woodcarvings cannot be seen at the present time.

Right next to the Villa Amerika and the Lapidarium, in the Na bojisti 14, is one of the most important pubs in Prague, **U Kalicha**. Although it lies in an unassuming street in the New Town district of Prague, the buses parked in front of it are a sign that something special is going on here. And it is, too.

This is where the Prague author Jaroslav Hašek frequently used to drink, and his novel *The Good Soldier Schwejk* has made the "Chalice" famous. "When the war's over, come and visit me. You'll find me in the Chalice every evening at six," Schwejk says to his friend. Today, the pub has become a place of pilgrimage. The walls are covered with paintings and quotes, the waiters dressed up – all is familiar from the book and the film. However, there is one drawback to U Kalicha – the countless coachloads of tourists who arrive at

Fleků – n absolute ust for ose who dmire zech beer.

lunchtime. If, by the way, you want to continue on the trail of the Bohemian (in both senses) Jaroslav Hašek, you can visit the house where he was born. It is also in the New Town, on the Školská which branches off from the Vodičkova.

Another famous author who lived in the New Town is the Austrian Franz Werfel, who lived near the City Park and whose father owned a glove factory in the Opletalova behind the Hotel Esplanade. The famous **Café Arco** was the meeting place of the "Arconauts" – Franz Werfel, the "roving reporter" Egon Erwin Kisch, Franz Kafka, Max Brod and the editors of the German language paper *Prager Zeitung*. It lies in the northern half of the New Town, in the Hybernská 16. Werfel, Kisch, Kafka and Brod once dominated the intellectual life of Prague, but those times are now finally over and the Café Arco is no more than a sad relict. Franz

Kafka's grave is in the **New Jewish Cemetrey** *(Židovské hřbitovy)*, and a visit can be easily combined with a tour of the New Town. You get out at the Metro station Line A, *Želivského*. The route to the grave is marked on a tablet near the entrance to the cemetery. However, the cemetery is closed to visitors on Saturdays.

The **main railway station** *(Hlavní nádraží)*. is worth taking a look. It is originally an Art Nouveau building, with a large and spacious roofed entrance hall. The combination has on the whole been a success. Not far from the station, but already in the district of **Vinohrady** (which gets its name from the vineyards that once thrived here), is the **Riegrovy sady**, a large and well laid out park along the slopes of the hill. This offers a beautiful view across the whole city right up to the Hradčany. A big beer garden also belongs to the park. It is not overflowing with tourists, and on most evenings a small brass band plays there.

A little out of the way, no longer in Nové město but in the district of Žižkov, is the **National Monument** *(Národní památník)*, an immense granite-faced cube with the "Tomb of the Unknown Soldier". In front of it stands one of the biggest equestrian statues in the world – the monument to the Hussite leader Jan Žižka. The heroic Žižka, whose military experience included fighting for the British at Agincourt in 1415, was chosen leader of the popular party after the outbreak at Prague on 30 July 1419. Having conquered emperor Sigismund's army and captured Prague after the battle of Vítkov Hill in 1421, he erected the fortress at Tabor, from which the radical Taborites get their name. Despite losing both eyes in battle, Žižka went on to secure religious liberty for the Hussites.

Only the Stalin monument, which had dominated the Letná hill until it was demolished in 1963, was bigger than that of Žižka.

Left, just like in the Schwejk novel – *Schwejkomania* in the U Kalicha. Right, Jan Palach is remembered on Wenceslas Square.

ZDAR TOBĚ PRAHO VZDORUJ ČASU

JAK ODOLALAS VĚKY BOUŘÍM VŠEM!

To the northeast of the Letná Hill is an extensive area set aside for recreational and cultural pursuits. It includes the **Tree Garden** (*Stromovka*), a large park complete with Neo-gothic hunting lodge which was first conceived in medieval times. The artificial stream is fed from a tunnel leading from the Moldau; regarded as a miracle of engineering when it was created in the 16th century, the system still functions today.

The adjacent **Exhibition Grounds** (*Výstaviště*) were opened in 1891, and from 1920 they were the site of the annual Prague sample fairs. In 1955 the whole area was converted into a recreation park providing a variety of cultural, sports and entertainment activities. Of particular architectural interest is the **Palace of Congress**, an iron girder construction from1891. Built in the same year, the **Pavilion of the Capital Prague** today houses the impressive **Lapidarium** of the National Museum, a collection of architectural fragments dating from the 11th to the 19th centuries.

In the round **Lipany Pavilion**, the visitor can marvel at L. Marod's panorama painting *Battle of Lipany*, the battle in which combined Catholic and Utraquist forces defeated the radical Hussites in 1434. The **Planetarium**, built between 1960 and 1962, and provides regular demonstrations of the development of astronomy and the universe.

The **Sports Hall**, which can hold as many as 18,500 spectators, grew out of the old pavilion which was used for machine building exhibitions; these days it is also used as a venue for concerts. The large **Swimming Stadium** was constructed in 1976. The **New Exhibition Pavilion** was designed for the Expo '58 held in Brussels, where it won a number of design prizes.

But the most special of all the attractions must be the so-called **Krizík Fountain**, built in 1891 and fully restored for its centenary. With its 50 pumps feeding almost 3,000 jets, there is nothing else quite like it in Europe. Every evening, 1,248 coloured underwater reflectors and 55 loudspeakers transform this 25-metre by 45-metre area of water into a symphony of syncrhonised acoustic and optical effects. The fountain can perform the entire repertoire of classical music.

This amazing contraption also comes into its own when used in conjunction with summer evening stage performances such as Shakespeare's *A Midsummer Night's Dream*: the ghosts, fairies, elfs and goblins from the Shakespearean magic circle come alive amidst the glittering dances of the fountains. So effective is the display that in 1991 the Czechoslovakian Music Foundation awarded this production the prize as the best musical performance of the year.

The performances begin according to the time of sunset, either at 9.30 or 10pm. The exhibition grounds can easily be reached by Metro, Line C, getting out at *Holešovice* or *Vltavská* station.

Preceding pages: the Municipal House; Pionkrál, mythical kings, in Vojanovy sady. Left, Zora Jandová as "Mother" in *A Midsummer Night's Dream*. Right, window cleaner outside the Hotel Forum.

Theatre in Prague has a long and venerable tradition. The spectrum is broader and enlivened by more original productions than many a "Westerner" imagines. It runs the gamut from *Laterna Magica*, a "theatre between dream and reality", via the the three theatres of the *National Theatre* with their programmes of top-class opera, plays and ballet, to the operetta productions of the *Hudební divadlo Karlín* theatre. With *Spejbl and Hurvínek* it crosses the boundary of puppet theatre for children; offers avant-garde samples of "silent theatre" through the accomplished mime theatre *Bránické divadlo pantomimy* and provides the additional choice of unusual attractions such as the *Komorní opera Praha*, *Cerné divadlo* (the Black Theatre), or the *Pražsky Komorní balet*.

In the outskirts of Prague there are little theatres and companies that often spring up quickly and for various reasons disappear just as fast. Every visitor to Prague can read in the daily national papers or in the monthly programme of the Prague Information Service (Na Příkopě 20), published in various languages, where and when such performances take place.

Laterna Magica is known all over the world as a synonym for perfect illusions, for a fascinating, minutely precise combination of theatre, film, mime and dance. All these disparate sections never have an independent effect, they always work as a synchronised whole. For 30 years the 60 or so members of the company have tried to make fantasy and poetry, comedy and tragedy real. By means of well-planned technical skill with projections, movable walls, stage props and actors, the plays move between light and shadow, uniting film and theatre, mime and dance into one extraordinary stage experience. At the centre of the spectacle is the visual impression, the dialogue between stage and screen. Every new production by this "theatre of light" is governed by a new key concept. Following on from well-established programmes such as *Magic Circus*, performances like *Odysseus* (adapted from Homer) or the ballet *Minotauros* with a libretto by the Swiss author Friedrich Dürrenmatt have achieved worldwide recognition.

What do the three theatres of the National Theatre – the **National Theatre** (*Národní divadlo)* itself, the **Smetana Theatre** *(Smetanovo divadlo)*, and the **New Theatre** *(Nová scéna)* – have to offer opera and ballet lovers? The National Theatre is the largest theatrical institution in Prague and employs 2,000 people. Its programme includes not only drama and opera but also works by famous native composers.

The famous Czech composer Bedřich Smetana is particularly well represented at present. Visitors to Prague can admire the whole range of his romantic operas: *Libuše* (about the mythical princess who founded Prague); the village opera *The Kiss (Hubička)*; *Dalibor*, set in a world of medieval knights; the world-famous folk opera *The Bartered Bride (Prodaná nevěsta)*; *The Secret (Tajemství)*, a comic opera dealing with a small town feud; *The Devil's Wall (Čertova stěna)*, another comic opera which describes the quarrel of the devil with a dishonest hermit; the society operetta *Two Widows (Dvě vdovy)*; and finally, the opera *The Brandenburgs in Bohemia (Braniboři v Čechách)*, which celebrated its first success in 1866 shortly before the battle of Sadowa and has been in the repertoire ever since.

Antonín Dvořák, another distinct Czech composer who received recognition from the musical establishment before Smetana, still has his opera *Rusalka* and other works performed. Bohuslav Martinů has also now become world-famous, but at present only his opera *Ariadne* is in the repertoire.

Admirers of contemporary music will mostly find their tastes represented by the works of Leoš Janáček. In this case, however, Prague does at least have the privilege of authentic performances, including *Jenufa (Její pastorkyně)*, an opera as tragic as *Katja Kabanova–(Kát'a Kabanová)*, which is based on Ostrovskij's *The Storm*; the poetical and philosophical work that deals specifically with nature and the human race, *The Cunning Little Vixen (Příběhy Lišky Bystroušky)*; and the well-crafted criticism of

Left, between dream and reality – a performance by Laterna Magica. Above, the seats in the National Theatre.

Zábradlí

PANTOMIMA
LADISLAVA FIALKY

bourgeois life, *The Excursions of Mr Brouček (Vyleti pana Broučka).*

The chamber opera company **Komorní opera Praha** uses younger voices, and includes Mozart's *Cosi fan tutte* in its repertoire. The company performs in the Palace of Culture or in the *Klicperovo divadlo.*

The *Černé divadlo,* Jiří Srnec's famous **Black Theatre**, is based in Prague but rarely performs here, as the company is mostly on tour abroad. Equally rare are performances by the **Prague Chamber Ballet**, *Pražsky Komorní balet*, directed by P. Smok. The company has no theatre of its own and is also on tour abroad more often than it performs in Prague.

The **Karlín Music Theatre** (*Hudební divadlo Karlín*), stages operettas, often lavish productions including *Die Feldermaus, The Land of Smiles* and even *My Fair Lady*.

It was in the theatre **Divadlo Na zábradlí** that Václav Havel made his first steps on the "stage that means the world". Here he worked as a scene-shifter and dramaturge and it is here that almost all his plays were first performed. Another great figure of Czech Theatre, whose name is also inextricably linked with this institution, was the mime artist Ladislav Fialka. His limitless imagination and inimitable style inspired a whole generation of mime and made Prague a world centre of this form of theatre.

Mime has developed further in Prague in recent years. In 1981, the **Branik Mime Theatre** (*Bránické divadlo pantomimy*) was formed, in which various groups such as CVOC or MIMTRIO appear.

Špejbl and Hurvínek is a **puppet theatre**, a perfect entertainment choice, for children as well as adults. The theatre has become well-known in the West through TV performances. It is the only theatre company with two main comic figures: the narrow-minded father Špejbl, "born" in 1920, and his son Hurvínek, some six years "younger", who is more exuberant, but also more intelligent. If you happen to be visiting Prague with children, don't miss an opportunity to see a performance.

Left, Ladislav Fialka, a classic performer of mime. Above, in front of the Novo divadlo – the New Theatre.

Today, once again, the **Estates Theatre** is not just an architectural jewel for visitors to come and marvel at. After eight years of being renovated, it was re-opened in 1991 to the strains of Mozart's *Don Giovanni*. Just over 200 years previously, on 29 October 1787, the opera had its world première on this very same stage.

For Mozart this theatre is linked with one of the happiest periods of his entire musical career. The theatre itself has a varied history.

The representative venue was innaugurated as the Count Nostitz Theatre in 1783 with a performance of Lessing's *Emilia Galotti*. In 1798 is was bought by the Czech estates who changed its name to the National Theatre. The Czech national anthem *Kde domov muj?* (where is my homeland?) resounded from the stage for the first time in 1834, during a performance of the comedy *Fidlovačka* (Spring Festival) by Josej Kajetán Tyl. The name was changed again to Tyl Theatre, when the Communists came to power in 1948. Hopefully, the fourth change will be the last.

THE FORMER PRAGUE GHETTO

According to records, the first Jewish community was founded in Prague in 1091. The Jews' relationship with the authorities always alternated between acceptance and hatred. During the Age of Enlightenment, the former Ghetto was renamed "Josephstown" *(Josefov)*, in honour of the reforming Habsburg emperor Joseph II. Later it became the fifth district of Prague. After the clearance programme of 1890–1910, decided upon and carried out by the Prague city administration, only the Jewish Town Hall, six synagogues and the old Jewish cemetery remained. These are now administered by the National Judaic Museum. The National Socialists wanted to create a "museum of the extinct Jewish race", but after the liberation, it was founded as the home of the largest collection of sacred Jewish artifacts in Europe.

The **Jewish Town Hall**, Maiselova 18, was constructed in 1586 in a Renaissance style by Pankratius Roder. The alterations in 1756 were the work of Josef Schlesinger. Between 1900 and 1910 the southern part was added.

The **Old New Synagogue**, the oldest remaining synagogue in Europe, is an unparalleled example of a medieval two-aisled synagogue. Services are still held here. On the outside the building has a plain, rectangular shape, a high saddle roof and a Late Gothic brick gable. The outer walls with their narrow pointed windows are strengthened by buttresses. The low annexes surrounding the main building served as entrance hall to the synagogue and as the women's aisle. The consoles, the capitals of the pillars and the vaulting are all richly decorated with relief ornamentation and plant motifs.

In the centre of the main aisle, between the two pillars, is the Almemor with its lectern for reading the Tora, separated from the rest of the interior by a Gothic screen, decorated with asses' back motifs. In the middle of the east wall is the Tora shrine, formed of two Renaissance pillars on consoles, with a triangular tympanum.

The **High Synagogue** was originally part of the Jewish Town Hall built by Pankratius Roder, and was a functional part of it. However, in 1883, the entrance to the Town Hall was closed. The hall of the synagogue is almost square, lit by numerous high windows, and gives a very worldly impression. The walls in the lower room are divided into three by flat pilasters, a device which echoes the crescent vaulting and the arrangement of the windows in the north wall. The central vaulting with its rich stucco decoration, which mirrors the profile effect of Gothic rib vaulting, shows how Renaissance forms adapted to Late Gothic taste. In the lower room

Left, the pinnacles of the Old New Synagogue. Right, the Jewish Town Hall in the Maiselova.

MAISELOVA

there is also a permanent exhibition of sacred textiles.

The **Maisel Synagogue**, founded by Mordecai Maisel, leader of the community in the old Jewish town, as a family synagogue in 1590–92, was built by Joseph Wahl and Judah Goldschmied in a Renaissance style, altered to a Neogothic style in 1893–1905 by Professor A. Grotte. A permanent exhibition of synagogue silver has had its home here since 1965.

The **Spanish Synagogue**, Dušní 12, was designed in 1868 by V.I. Ullmann. It has a square ground plan and a mighty dome covers the central hall. Metal constructions, which open out into the main aisle, lie on three sides. The marvellous decoration of the interior earned this synagogue the name "Spanish Synagogue".

Services were held in the **Cells Synagogue** until 1939. This is a baroque building with a longish hall and barrel vaulting. In 1694 it was built to replace the little "cells", buildings which served as houses of prayer and classrooms. The building has two rows of round arched windows in the south wall, facing the cemetery. The walls are divided by pilasters which support the roof beams under a projecting sill. The synagogue is now used to exhibit old Hebrew manuscripts and printed works.

The **Pinkas Synagogue** is architecturally a very beautiful Renaissance building. It came into being in 1535 in a specially adapted private house belonging to the leading ghetto family of Horowitz. In 1625 it was rebuilt in a Late Renaissance style by Judah Goldschmied and extended by the addition of a women's gallery, a vestibule and a meeting hall. Since 1958, the synagogue has been a memorial to the 77,297 Jewish victims of the Holocaust who came from Bohemia and Moravia. The inscriptions on the side walls list

Inside the Old New Synagogue.

name, date of birth and date of deportation to the extermination camps, in alphabetical order, for each victim.

The **Old Jewish Cemetery** is reckoned to be one of the 10 most interesting sights in the world. It came into being in the 15th century, when pieces of land on the northwest edge of the ghetto were bought up. Burials continued here until 1787. The number of graves is much greater than the remaining 12,000 gravestones. Existing graves had to be covered with earth to form new graves, which is what has caused the hilly landscape of the cemetery and the characteristic layering of several graves one above the other.

The inscriptions on the stones give the name of the deceased, the father's name – in the case of women, the husband's name as well – the date of death and the date of the funeral. The majority of the inscriptions consist of poetic texts expressing grief and mourning. The reliefs portray the name of the deceased, the profession, the connections with priestly or Levite families (blessing hands) or membership of the tribe of Israel (grapes).

The oldest monument is the tombstone of the poet Avigdor Karo, dating from 1439. Also buried here in 1609 were the famous scholar and supposed creator of the Golem, Jehuda Löw; in 1601 the leader of the Jewish community, Mordecai Maisel; in 1613 the scholar and astronomer David Gans; in 1655 the scholar Schelomo Delmedigo; in 1735 David Oppenheim. A splendid tomb was built in 1628 by Wallenstein's financier, Jakob Bassevi von Treuenburg, specially for his wife Hendele.

In the Neo-romanesque **House of Ceremonies** there is an interesting exhibition of children's paintings and drawings from the concentration camp of Theresienstadt.

e kosher staurant the ewish own Hall.

HOLOCAUST IN PRAGUE

The Prague Ghetto is one of the oldest in Europe, dating back to the 11th century and possibly earlier. By the 17th century it was a flourishing community and a focal point for Jewish culture in Central Europe.

The Jewish community in Prague was allowed to develop freely after 1848 (when the laws segregating the Jews were finally repealed), until it was almost completely destroyed in the Holocaust following the occupation of Bohemia and Moravia on 15 March 1939 by the Nazis. Today there are some 1,200 members of the Jewish community left. There are also about another 1,500 people of Jewish descent living in Prague.

The Nazis systematically planned and carried out the "Final Solution" to the "Jewish problem". The Jews were first of all forced to register, then excluded from economic activity, then physically separated from the rest of the community. They were marked, insulted, sentenced to perpetual poverty, and evicted from their homes. They were deported to concentration camps, starved and tortured physically and mentally. For those who could no longer work, the death sentence was inescapable.

In the "Protectorate of Bohemia and Moravia" the Jews suffered everything, apart from the "Reichskristallnacht", that had been happening to the Jews in Germany from 1933 on. Immediately after the occupation of Bohemia and Moravia by Hitler's army, the "Nuremberg Laws", passed in 1935 "to protect the German race", were declared to be in force retrospectively. These laws had deprived German Jews of their citizenship and turned them into "subjects of the state". The same now happened to Czech Jews. One persecution of the Jews followed another – culminating in the mass deportations of 1941.

The first five transports, each with 1,000 people mainly from the Jewish intelligentsia, – doctors, artists, lawyers – were deported to the so-called Litzmannstadt Ghetto in Lodz, which the Nazis themselves designated a starvation camp. A month later the old fortress of Terezín (Theresienstadt) in Bohemia was declared a Jewish ghetto, and received further deportations not only from Prague, Bohemia and Moravia, but from all over Europe. Theresienstadt was not itself an extermination camp, but from it the Jews were sent on for so-called selection and thence to the gas chambers of Auschwitz.

However, even in times of the greatest debasement of humanity there remained artists in Prague who had enough courage at least to lessen the impact of the humiliations they suffered. Authors wrote under the names of their friends; cultural afternoons to discuss contemporary poetry were held in private apartments. An amateur drama group led by the famous Shakespeare translator Erik Saudek put on plays. The author Norbert Frýd's verses, *A Horse Decorated with Flowers*, an introduction to the alphabet, were set to music by the composer Karel Reiner and performed by both for the children of the orphanage. In the orphanage in Košiře Glucks, the opera *The Deceived Kadi* was also performed.

Left, tombstone on the Old Jewish Town Hall. Above, in the concentration camp.

VYŠEHRAD

The Vltava flows down from the Bohemian forest and reaches Prague by the rock of the Vyšehrad. This is where, according to legend, the age of myths ended, and the rule of the wise women, skilled in magic, was replaced by the rule of men. The marriage of the Princess Libuše to the farmer Přemysl brought this change about, and their successors ruled over the weal of the Czech people up until 1306 of our era. It was here on this rock, where the couple are supposed to have lived in a magnificent palace, that Libuše had her great vision in which she prophesied the future greatness and glory of the new capital.

Not until the 19th century did the Vyšehrad enter into the newly revived Czech national consciousness. The legend was embroidered with much imagination and the Vyšehrad became once more the seat of Libuše and the cradle of Czech history. Many artists, poets and painters, musicians and sculptors, historians and architects worked on the site. In this way a sort of memorial was achieved which in its own special way says something about this nation living in the centre of Europe – Slavs, surrounded by German tribes, involved in many complex relationships with these neighbours, and yet different in character and speech.

The oldest building on the Vyšehrad is **St Martin's Rotunda**, a Romanesque church constructed some time after the year 1000. It is one of the oldest Christian churches in the country. There are similar rotundas in various places in modern Prague, for instance the **Holy Cross Rotunda** in the Old Town or **St Longinus** in the New. They are all that now remains of the cores of former individual settlements.

More is known about the **Church of St Peter and St Paul**, which the visitor can see in its Neo-gothic form, dating from 1885 to 1887. At the present time archaeologists are examining the walls of its predecessor on this site. In earlier times the Vyšehrad was the goal of pious pilgrims. Here, in St Peter and Paul's church, the votive tablet popularly known as the **Madonna of the Rains** was kept. It is now in the collection of St George's monastery in the Hradčany.

The redundant fortress was demolished in the 19th century, for the Vyšehrad had long lost its strategic importance. A centre for the Czech people was created. The heart of these patriotic efforts during the 1870s was the creation of the **Slavín**, a special cemetery. It was designed by Anton Wiehl and later completed by the addition of the tomb of honour at the end of the main avenue, with its ornamental sculptures by Josef Mauder.

Preceding pages: a dog and his master in the Letná Park. Left, portal of St Martin's Rotunda.

Many of the graves are still regularly decorated with flowers. No "soldiers" or "heroes" are buried here, only poets, musicians and artists. The works of these artists live on in the memory of the nation. The most popular are the two best-known Czech composers, Bedřich Smetana and Antonín Dvořák. Smetana's *Bartered Bride* alone has been performed in Prague more than five thousand times. However, these two composers are not the only musicians buried there. There are also great performers such as violinist Jan Kubelík or the virtuoso Josef Slavík, very famous in his day.

Also buried here is the author of the stories of life in the Malá Strana, Jan Neruda. His stories are of the world of the lower middle classes, living over there in the Malá Strana in the shadow of the palaces. They feature old women and their books of dreams, moonstruck students, grumbling caretakers, and a number of curious characters from the backyards. Also of Neruda's generation and buried in the Slavín are Svatopluk Čech, Jaroslav Vrchlický and Karel Hynek Mácha, whose *May Poem* is known to every Bohemian.

The visual artists and painters are represented by Mikolás Aleš, Josef Myslbek and Jan Štursa among others, and more modern times by the author Karel Čapek, or by Alphonse Mucha, who is best known for his Art Nouveau posters.

The Vyšehrad can easily be reached from the Metro station *Vyšehrad*, passing the gigantic Palace of Culture. However, you can also get to it on foot along a pleasant path from the Slavojova Cikova or directly from the Vltava bank through the thickly wooded park to the castle. The remains of the old fortifications provide the visitor with a beautiful view of the city and the Vltava.

Peter and : Paul on e ⁄šehrad.

Excavations have confirmed that, in the area of present-day Czech Republic, glass was known and used in the form of bead necklaces and, later, bracelets. In the early Middle Ages the first blown glass was used for drinking goblets, and at the same time windows, glass panes and wall mosaics were manufactured. During the reign of Charles IV, in 1370, the massive mosaic on the south portal of St Vitus' Cathedral in Prague was made. It shows scenes from the Last Judgement. The splendid windows of St Bartholomew's Church in Kolín also date from around 1380. Glass for everyday use was, as in other Central European countries, mostly in the form of bottles and phials, greenish or brownish in colour.

By the beginning of the 15th century there is evidence for eight glassworks in Bohemia, five in Moravia and another eight in Silesia, which at that time was part of the Bohemian kingdom. The glassworks in Chribská, which still exists today, had already been mentioned in 1427, and had considerable influence on the development of the Bohemian glass industry. By the 16th century, the rustic glass made in the forests no longer satisfied the refined tastes of the humanist aristocracy who belonged to the court of the first Habsburg rulers. Following Venetian models, thin, refined glass was produced, in harmonious Renaissance styles. Towards the end of the century and after 1600 cylindrical tankards formed the ideal basis for the famous Bohemian enamelled glass.

From 1600 to 1610, at the Prague court of Rudolf II, the jeweller Caspar Lehmann (1563–1622) of Uelzen had experimented

with engraving glass. Lehmann was gem cutter to Rudolf II, and he adapted glass techniques of cutting gems with bronze and copper wheels. This was a new skill used for ornamenting glass, and engraved glass was proudly born.

In Bohemia the production of engraved glass did not begin until around 1680. From this time on, glass was exported in great quantities to all sorts of destinations. One example: between 1682 and 1721 Georg Kreibich from Kamecky Senov made a total of 30 business trips throughout Europe. Among the production managers of famous glassworks in north and northeast Bohemia and in the Bohemian forest the names of aristocratic families such as Kinsky or Harrach can be found.

The classic shapes of the Bohemian baroque goblet with its polished balustered foot, or the many-sided beaker of fine, thin glass, were intricately engraved with elaborate ornaments featuring flowers, garlands or grotesqueries.

After 1720, the available variety was enriched by the addition of gilded glass and black painted decoration in the style of Ignaz Preissler (1676–1741). Engraved and cut glasses from Bohemia were famous, as was the Bohemian chandelier with pendants of cut crystal glass. Even in the 18th century these chandeliers were being exported to the courts of France and Tzarist Russia.

After some stagnation towards the end of the 18th century, milky Bohemian glass with painted, colourful enamel decoration became tremendously popular. Contrasting trends that developed around 1800 provided plain, finely cut and engraved Empire and Biedermeier style glasses.

New success came from 1820 onwards through the thick coloured glass discovered by Friedrich Egerman: black hyalith (often with gold Chinoiserie decoration), agate glass, lithyalin (which resembled semi-precious stones) and new uses for ruby glass and opaque white overlay glass, both carved and enamelled, were taken into production.

The fame of Bohemian glass was increased in the 19th century by mirror and plate glass and also by glass coral. The latter is still produced today in Jablonec nad nisou, now under the company name Jablonex.

Bohemian glass entered the 20th century with names such as Loetz, Lobmayer, Jeykal and others, with imaginative shapes formed by Art Nouveau, with surprising colours and metallic effects. Apart from its usual domestic and technical functions, modern glass has become an art form. The Arts and Crafts School in Prague produced a particularly excellent generation of artists just after World War II. They had great success in many important exhibitions in the 1950s and '60s.

Left, glass window in St Vitus' Cathedral. Above, the famous Bohemian crystal.

LORETO AND NOVÝ SVĚT

If you walk up from Hradčany Square toward the Strahov monastery, now the Museum of National Literature, you can hear, every hour on the hour, a delicate tune played by bells as you walk between the palaces.

Many years ago, during the plague, there lived in Prague a mother with her children. One child after another fell sick, and with the last few silver coins that she had left she paid for the church bells to be rung whenever a child died. The poor widow had to give up coin after coin, all her children fell victim to the plague. In the end, after all the children were dead, she herself fell ill and died. However, there was now no-one to have the bells rung for her. Then, all of a sudden, all the bells of Loreto rang out and played the tune of a famous hymn to Mary. So it has remained right up to the present day.

This little anecdote from Prague gives a clear indication of the importance that this shrine has for many people. It is not merely of historic and artistic importance, it is still considered a place of pilgrimage.

The Santa Casa: In the mid-13th century the armies of Islam invaded and conquered the Holy Land. At that time two brothers were priors of the Franciscan monasteries in Haifa and Nazareth. When they fled, they probably took all that was most precious to them along. According to legend, they removed the Santa Casa stone by stone, and eventually rebuilt it near Renecati, now Loreto in Italy. Later the house, visited by many pilgrims on their way to Rome, was decorated with rich marble reliefs, and the many copies also show this particular ornament. When the Catholic Habsburgs tried during the

The baroque tower of the Loreto in Prague.

Counter-Reformation to convert their Hussite subjects back to the "true faith", they used the pious legend to serve their cause. They had replicas of the Santa Casa built throughout the land. The best known and most attractive of these is the **Loreto of Prague**. It stands on the Loretánske nám. and was founded by the Blessed Catherine of Lobkovic, who laid the foundation stone on 3 June 1626.

Unlike the simple original, the shrine became across the centuries an entire complex consisting of various buildings with a chapel, cloisters which were several storeys high, and the church of the Nativity. Dominating the group is the early baroque tower, into which the carillon which chimes every hour was built in 1694.

Just as in Loreto, the shrine's outer walls are decorated with Renaissance reliefs. The interior also strictly follows the Italian model. As a result, you can see in the Prague Loreto a small, bare building which is very probably the copy of a house in Palestine, and in which the **Loreto Madonna** is honoured. Dressed in a long cloak, she carries the infant Jesus in her arms.

The two-storey cloisters surrounding two courtyards were the work of the Bavarian baroque architect, Kilian Ignaz Dientzenhofer. The paintings in these cloisters have been heavily overrestored and you can hardly see any traces of their original beauty. However, the wealth of the poetic images in the supplications to Mary is all the more impressive: "Tower of David", "Gate of Heaven", and again and again "Oroduj za nas" – pray for us.

Between the portal and the Santa Casa you can see the **Church of the Nativity**, a room decorated by notable artists, which was consecrated exactly 111 years after the laying of the foundation stone, on 7 June 1737.

yllic peace Nový ět.

As at many other places of pilgrimage, successive pilgrims here have also given votive gifts to the treasury as a sign of thanksgiving. And the Loreto's main attraction is the indeed the **Treasure Chamber**. The gifts of the Bohemian nobility were commissioned from the most notable goldsmiths of the time and are some of the most valuable works of art among liturgical objects in Central Europe. The most remarkable is the **diamond monstrance**, which was a legacy of Ludmilla Eva Franziska of Kolowrat, who left everything to the Madonna of Loreto. The monstrance was made in 1699 by Baptist Kanischbauer and Matthias Stegner of Vienna. The monstrance, studded with 6,222 diamonds, sends out its rays like a sun. It is almost 3 ft (1 metre) high and weighs more than 26 pounds (12 kg). Right next door to the Prague Loreto, the restaurant *U Lorety* is a pleasant example of a Prague garden restaurant.

Palais Czernin: If you leave the Loreto and walk on right up to the square, you are almost thunderstruck by the truly massive facade of the **Palais Czernin**, an incredible counterweight to the light building surrounding the Santa Casa, which, seen from this point, almost seems to cower. Twenty-nine half-pillars are here lined up one after the other and their height (more than two storeys) defines the palace facade, over 490 ft (150 metres) long.

In 1666, Humprecht Johann, Count of Czernin, bought the land, and work started on the palace straight away, under the direction of Francesco Caratti. In 1673, the Emperor Leopold I came to Prague and demanded to see the building about which there was so much talk in distant Vienna. It did indeed seem as if the Count, who had not received the imperial favour he expected, was building his own palatial residence out of pique. At any rate, the emperor was

Window gallery in Nový Svět.

highly displeased when the Count announced that really it was nothing but a big barn and of course he wasn't going to leave the present wooden doors in, but was going to replace them with bronze. "For a barn, those wooden doors are quite good enough," the emperor retorted.

The Czernin were an old Bohemian family and their members had excelled time after time in the service of the Bohemian crown. The house in Prague was to become a "Monumentum Czernin", but fate was no longer kind to the building. Construction work continued for several generations, until at length financial collapse put a stop to the project. It was partially destroyed during the sieges and wars, and in 1779 the heavily damaged building was to be sold; no buyer came forward, however. During the Napoleonic Wars, the building was a military hospital, in 1851 the state bought parts of it and turned it into a barracks. In 1929, the authorities of the young Czechoslovak Republic had the palace renovated and made into the Foreign Ministry.

In the arcades opposite the Palais Czernin more mundane things are to be found. If you like dark beer, here in **The Black Bull** (*U Černého vola*) you will find the good and strong (12°) Velkopopovicky kozel.

Nový Svět: Below the gardens of the Palais Czernin runs a alley which belongs to the old settlement in front of the castle. In the middle of this former poor quarter is the **New World**, *Nový Svět*, which now draws many an artist and intellectual. The houses all have names, many have a house sign with the adjective "golden". This is something typical of Prague, and so the houses are called "The Golden Leg", "The Golden Star", "The Golden Pear". In **The Golden Pear**, *U zlaté hrusky*, is a romantic wine bar.

**e "Golden
ar"
ern.**

THE ROYAL WAY

The last coronation in Prague took place on 7 September 1836. Ferdinand I, the Austrian Emperor, became the last crowned King of Bohemia. In 1848, Ferdinand I abdicated in favour of his nephew Franz Joseph, who, however, did not allow himself to be crowned. Charles, the last Habsburg emperor who reigned from 1916 to 1918, had himself crowned with full ceremony as King of Hungary – and in the middle of World War I too – but not in Prague.

People in Prague took umbrage at this behaviour by the Habsburgs in Vienna; after all, ever since the imperial seat had been moved from Prague to Vienna, all of them had at least come here to be crowned king or queen of Bohemia. And all of them had followed the route that is rightly called the Royal Way: from the **Powder Tower** to the **Celetná**, the **Karlova**, the **Charles Bridge**, the **Nerudova**, across the two big squares, the **Old Town** and the **Malá Strana** squares – even today, this route is designated the **Royal Way** by those who are officially responsible for the monuments of Prague.

Only with difficulty can one imagine the carriages with their four or six pairs of horses, the riders and runners proceeding through the streets. The buildings were for a long time subject to damage from the vibration of modern traffic, until they were turned into pedestrian precincts. Now the streets have quite a different appearance from how the Royal Way must have looked. Today they are dominated by shops and restaurants, by tourists and Prague families out for a stroll. What did it look like in earlier years? What did the people of Prague see when they lined the narrow streets of the Royal Way?

The coronation procession: Today you can follow one of these processions step by step and discover exactly where each salutation took place, how long the procession stopped, where the cheering crowds gathered. According to one contemporary account: "On both sides many thousand people of both sexes, all filled with joy, who continually cried 'Long live Maria Theresa, our most gracious Queen!' to give utterance to their rejoicing." On 29 April, 1743, on account of the "sudden inclement and windy weather", the queen had to travel in a closed carriage, drawn by six dark brown Neapolitan mares, strong and handsome creatures. To the left of Maria Theresa sat her husband Francis Stephen of Lorraine, soon to become Emperor Francis I.

In those days the Horse Gate at the end of the Horse Market (now Wenceslas Square) was still standing. A great tent had been erected in front of it, into which the queen had withdrawn

receding ages: cade of e Palais zernin. **eft**, Maria heresa. **ight**, Charles IV" the ělník estival.

for a short while. In the meantime the procession formed up, all 22 groups of it. All were on horseback, in new uniforms, the ladies in splendid coaches, accompanied by musical bands. After crossing the New Town in a wide curve, via Charles Square and Na příkopě, the procession came to the Powder Tower and the Old Town. From here on it went through the Celetná, passing what was once the most exclusive hotel in the Old Town of Prague, the "Golden Angel", and the Týn Church with the courtyard behind it, then a hostel for travelling merchants passing through.

When the royal procession reached this focal point, they assembled in front of the Týn Church for the deputation of all four faculties of *Alma Mater Pragensis*, and each nobleman or woman to be greeted with a fine, well-turned speech – in Latin, of course.

Once the ceremonies of the Old Town Square were over, the procession passed the magnificent houses, several storeys high, that lined the route before entering the narrow Jesuit Alley (Charles Alley today), which winds around a number of bends till it meets the Crusader Knights Square.

The last third of the way passed the Clementinum and St Saviour's Church to the Old Town Bridge Tower. In spite of its slender form, the tower was originally designed for defensive purposes, proving its worth in 1648 when the Swedes spent two weeks vainly trying to conquer the Old Town. The west side of the tower was destroyed, but on the Old Town side the gallery of sculptures has survived. Charles IV still wears the imperial crown and the shield at his side bears the imperial eagle. At his side is the figure of his son Wenceslas IV wearing the crown of his saintly patron and bearing the imperial regalia in his hand. Between them, standing on a kind of model of the bridge, is St Vitus,

Family tree of Charles IV in Karlštejn Castle.

patron saint of the city. The heraldic shields, lined up as if in greeting, display the crests of countries that were ruled by the king of Bohemia.

The **Charles Bridge** was not always the avenue of saints' statues who seem to be offering sound moral advice to those passing by. The coronation procession made its way over the bridge to the Malá Strana, that part of Prague that is so different from the narrow streets of the Old Town, and then entered the **Bridge Alley** between the Malá Strana Bridge Towers. The Alley is short and quite broad, and above, high above the proceedings, rises the dome of the church of St Nicholas. The procession passed along the south side of the church, the symbol of the Malá Strana and of the Prague baroque style. Here the last part of the Royal Way began, the steep street that is now the **Nerudova**, which demanded all the attention of coach drivers and riders. It was particularly important to manage the great curve into Hradčany Square smoothly and to keep the horses moving at an even pace.

The procession passed many houses with their delightful house names; the "Three Violins" or the "Two Suns", their signs above their gates visible for quite a distance. And yet two palaces have squeezed even into this stretch of road, that has been built on for over a thousand years. The first was commissioned in 1715 by the Morzin family from the famous architect Santini, the second built by Matthias Braun for the Kolowrat family.

From here, it took the royal procession only a few more steps to reach Hradčany. This is where the heir was presented to the people, a ceremony that lent the occasion something of the spirit of an election. But the candidate's succession had long been established by the laws of inheritance.

...e last ...rve before ...e Prague ...astle.

Pubs And Coffee Houses

The "coffee house", that great Prague institution from the years before and between the two world wars, no longer exists. It used to be a place which had the latest newspapers, where the waiters were more like friends and confidants, which was patronised by the "great" ladies of contemporary society and where the ambience was created by the flair of artists and journalists.

There were big, ostentatious coffee houses in the city centre – here the waiters, according to the writer Jaroslav Seifert, at any rate, went twice a day to the barber's to let themselves be shaved. Every coffee house had its "own" clientele – actors in the *Slavia*, Kafka, Kisch and the "Arconaut" circle in the *Café Arco*. Here couples met and held hands – over there was the meeting place of the demimonde.

The coffee – and this still hasn't changed even today – was well-known to be awful, and you paid your two crowns for it more or less as an entrance fee. In winter you came to get warm and to save on heating costs. In summer you came "for the thick tobacco smoke" – according to Seifert. For many people, the coffee houses were a way of life.

Even if the glories of times long past are no more to be found here, and the "Bohemian" lifestyle seems to be over – the coffee houses of Prague are still worth a visit. Especially interesting are their interiors, which seem so nostalgic nowadays, with Art Nouveau decor such as that in the **Evropa** or the **Municipal House**.

The last of the large and important coffee houses in Prague was and remains the **Slavia**. Nowhere else, except in the Slavia, will you find such a cross-section of coffee house customers. They may not come as regularly as they used to, but the actors, artists, and singers still come to the Slavia from the opera and theatre opposite. And it can still happen that a slightly greying prima donna will hold court among her friends. The eternal, timeless artistic stereotype – smoking heavily, wearing a beret, clutching a manuscript – can also be seen in the Slavia.

With a bit of luck, he'll let you pay for the wine and in return tell you some of the inner secrets of literary life in Prague. But in the Slavia you can also find the old ladies, factory owners' widows from the First (the "Golden") Republic. However, they have become a rarity in the Slavia, and meeting such a lady is quite an occasion.

But young people also come in, from the nearby Conservatory, to drink coffee between lectures or to meet their friends. And in the evening the café is a meeting place for "yuppies" – if there could be said to be such a thing in Prague – well-dressed, with hairstyles

eceding ges and : in the é Slavia. ght, a ourite ce in mmer, U ků.

à la David Bowie and Benetton clothes. Next to them, in customary black, sit those from the opposite end of the fashion spectrum: punks with colourful, wild streaks in their hair and heavy army boots.

The waiters may no longer go to be shaved twice a day, but they haven't lost any of their pride in their profession. Be that as it may, it's hard to find the coffee house atmosphere during the tourist season, when apparently endless streams of tourists descend on Wenceslas Square. If you want to see where the young men meet, you should go to the **Evropa** – but only in winter. Whatever, the coffee houses will hopefully continue to be established institutions in Prague.

Pubs and beer: Many people in Prague can tell you tales of their grandfathers, who drank 20 or 30 tankards of beer a day, played cards in the pub and never so much as paid a visit to the gents' room. These tales are, of course, exaggerations. What remains is the love of beer. Yes, Čedok does offer "beer parties" in Prague – organised merry-making in the Interhotel. This has nothing to do with the life in the pubs of Prague.

A simple restaurant or pub doesn't have to be clean, as long as it serves the right beer, and whether it comes from Smichov, Pilsen (Plzeň) or Budweis (České Budějovice), Czechoslovakian beer is probably the finest in the world. But it has to be freshly pulled, the head mustn't collapse straight away, and the taps and pipes have to be regularly cleaned.

It is said that you get better beer in the **Koruna** in Wenceslas Square, which is nothing but a snack bar full of vending machines, than in many a "traditional" pub or restaurant. Where do they have especially good beer on tap? One of the best places is still the **Black Bull** (U Černého vola) up in Loreto. Here you'll

***Pub brawl*, painted by Josef Lada**

still find many Prague "characters", with broad middles and loud voices, who drink their beer with such speed that a stranger can hardly keep pace.

Or just try one of the pubs around the corner, most of which are packed full. What about the beer garden in Riegrovy sady, where a band plays on the weekends in the summertime and people come in from the surrounding streets to enjoy a little dancing?

Of course, you mustn't miss the **Chalice** (U Kalicha) with its mementos of *Schwejk* or the **U Fleků** with its beer garden, cabaret and live music. Several other venerable institutions which still survive include the **Golden Tiger** (U zlatého tygra) in which you may, if you're lucky, find one of the most famous Czech writers, and **St Thomas'** (U svatého Tomáse), with its old vaulted cellar. The **U Medvídků** and **U Labutě** are two traditional pubs less haunted by tourists.

Wine bars: The Malá Strana wine bars were once simple places, where one met, got something to eat – just a small snack – and whiled the night away over good Moravian wine. But here also one or two changes have taken place. Wine bars today are often very formal and expensive restaurants, with a plethora of white napkins and starched tablecloths. Only a few have profited by being changed into pretentious tourist traps.

However, if you don't want to eat in grand style – in the **Lobkovic Wine Bar**, for instance – but prefer just to drink a glass of wine, there are still some places to go; the **U Golema** and the **U Rudolfu** in the Maiselova, for instance, or the "Golden Frog", **U zlaté konvice**, in the Melantrichova. The most famous and popular wine bar in Malá Strana is still the "Patron", **U mecenáše**, in the Malá Strana Square.

ndow on
Café
avia.

FJI

Not Just Franz Kafka and Schwejk

The "Golden City" on the Vltava has always been a golden fertile ground for poetic language. In the early 20th century German and Czech had equal status in Prague, and each produced a world-famous novel: **Franz Kafka's** *Josef K.* and **Jaroslav Hašek's** *The Good Soldier Schwejk.*

In those times, marked by the spirit of the turn of the century and World War I, by literary anarchism and avant-garde, the Prague literary scene was formed, which, with an added dash of "Bohemian" lifestyle, was to culminate in the 1920s and '30s and even grow to rival Paris.

"Uncontrollable, in a certain sense" – this quote from Kafka about an unexpected love may perhaps point to a strange phenomenon: authors writing in Czech rarely seem to have discovered anything sinister or secretive when they wrote lovingly of the city on the Vltava; authors writing in German, however, had a tendency to imagine all kinds of dark secrets in their view of Prague and to prefer tragic subjects.

Nora Frýd, who was born a German Jew, Norbert Fried, and became part of the Czech literary scene under the Czech version of his name after a "wretched and blessed" period of his life spent in Theresienstadt, Auschwitz and Dachau, was asked about this phenomenon. This was his reply: "The German-speaking authors, Jews or non-Jews, often concentrated on the dark events in the varied history of Prague and imagined tormented personalities. Jews or non-Jews, they turned to the Golem and other legends about or based on Rabbi Löw, or to the secrets surrounding Rudolf II's cabinet of curiosities. The Jews of the '20s probably felt hints of their terrible fate in 1938, the German-speaking non-Jews perhaps had an inkling of their ill fortune in 1945. Those non-Jews and Jews who wrote in Czech, however, preferred to see in Prague only its golden attributes, Prague as *Matička* (Mama)." Another quote has come down from Kafka: "Mama has claws!"

Left, portrait of the "good soldier" Schwejk. Above, Franz Kafka and Felice Bauer.

There were several cafés and pubs in Prague in which, occasionally and temporarily, the Czech, Prague German, German-Jewish, Czech-Jewish or Sudeten German authors met. However, Max Brod adds a caveat about the most famous of these places: "The tales told about the Café Arco are untrue or at least highly exaggerated. Not until Werfel and his clique made the new Café Arco into their "local" did it amount to

anything." Well, the coffee house that Werfel, Kafka and Brod knew and frequented no longer exists.

Rather than searching for the places where the liquid intake crystalised the literary genius of Prague, it is probably more productive to let the city, indeed the combination of the landscape and the city, work on you as a whole. This is how the poetry school of the **Prague Circle** came into being, which, as Max Brod claimed, had no teachers and no curriculum. The *spiritus rector* added: "Unless, you might claim, Prague itself, the city, its people, its history, its beautiful surroundings, the forests and villages that we walked through on enthusiastic hikes, was both our teacher and our curriculum." No doubt this education in openness and tolerance made it possible for Max Brod to support the Czech Hašek as much as his friend Kafka. It's well-known that Kafka would never have attained his position as the classic author of modern literature if Brod, against the express wishes of the author, had not rescued and sorted Kafka's manuscripts, thus saving this important work and making sure that it spread from Prague to the whole world. He had a similar success with Hašek – i.e. he made him into a classic author of modern Czech literature by internationally promoting Hašek's novel *Schwejk*.

It seems incredible today, but then, in the 1920s, the Czech "good soldier" of World War I did not find favour with the Czech intelligentsia, who considered him to be immoral, improper and damaging to the Czech national image. Even so, Brod sent a dramatised version to Berlin to Erwin Piscator, under whom Bert Brecht was working at the time. The successful translation into German by Grete Reiner helped the play to succeed. She used a native idiom of Prague, the German dialect current in the Malá Strana, and so converted the play into German with great originality. The German translation was Hašek's first step on the way to international literary recognition. All over the world people were delighted by the cunning dog handler, who found his way out of the "historic situation" of World War I through "nonsensical behaviour and a mask of clownishness" (Radko Pytlík).

It must be said that the Czechs are no longer entirely happy about the success of this book. After all, it led to the internationally current prejudice that all Czechs are like Schwejk – or that, at the very least, there is a bit of Schwejk in every Czech.

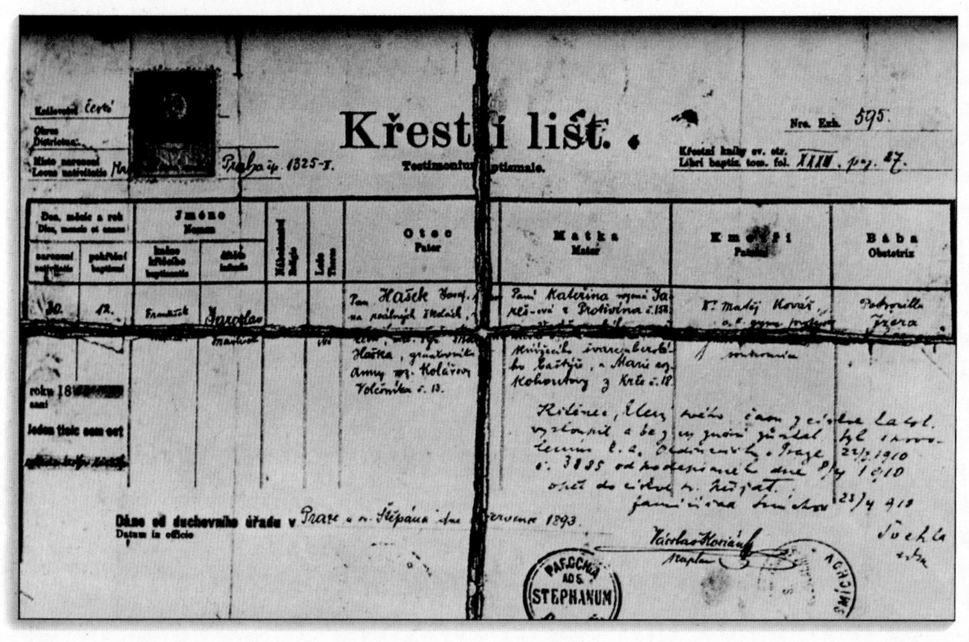

Křest[illegible] list.

Testimoniu[illegible]ptismale.

Nro. Exh. 595.

Křestní knihy sv. str. XXXV. pag. 27.
Libri baptis. tom. fol.

Otec / Pater: Pan Hašek Josef

Matka / Mater: Paní Kateřina

Kmotři / Patrini: Matěj Kovář

Bába / Obstetrix: Petronilla

Dáno od duchovního úřadu v Praze u sv. Štěpána ... 1893.
Datum in officio

Jaroslav Hašek was a Bohemian, in both senses of the word, of the first order. Within the space of an hour he could switch from complete concentration to apathy, he could work steadily towards a goal, yet within his lifespan of only 40 years he moved the goalposts several times at a moment's notice. He lived through it all, from anarchist gestures via plans for an imaginary Romanov on the Bohemian throne to work for the Communist Party, from the foundation of a "Party for gradual progress within the framework of the law" to alcoholic apathy. He lived through it all and included it all in his humorous and satirical masterpiece. Hašek was the literary creator of forgotten and of timely characters. His creative source was the pub, and he included many a quote from the bar. He immortalised the pub *U Kalicha* (The Chalice) together with the good soldier Schwejk, and thus left Prague with a meeting place for natives and visitors who want to combine alcoholic spirits with the spirit of literature.

However, modern literature in Prague meant not only Kafka and Hašek, Brod and the arcades and in the archways, on the bridges and in streets such as those between the Waldstein and the Malá Strana squares, in front of the statue of St Wenceslas and in the Hotel Evropa (then the Grand Hotel Schroubek).

The Czechs in Prague had in the 19th century already known their **Jan Neruda** and their **Božena Nemcová**. The *Tales of the Malá Strana* and the *Grandmother* had found their way, in translation, into German bookcases as well. The modern age of Czech literature developed with **Josef Čapek**, **Jaroslav Durich**, **František Halas**, **Vladimír

Čapek, Kisch and Nezval. Whole galaxies of literary stars, writing in both German and Czech, enlivened the city on the Vltava with their texts and their charisma, with their books and with their personalities. There may not have been a Montmartre as in Paris or even a Schwabing as in Munich, yet the writers filled the capital of the first Czechoslovak Republic with an atmosphere that can still be felt with surprising immediacy, under

Left, Jaroslav Hašek's birth certificate in the U Kalicha. Above, Karel Čapek is buried in the Slavín in the Vyšehrad.

Holan**, **Josef Hora**, **Josef Lada**, **František Langer**, **Marie Majerová**, **Ivan Olbracht**, **Vitězlav Nezval** and **Jaroslav Seifert**.

There's no doubt that there was little contact between the authors who wrote in German and those who wrote in Czech during the years between the wars. The unhappy nationalism of the 19th century continued unabated in post-1918 Czechoslovakia. Of course individual authors built literary and political bridges. Max Brod could still speak friendly words in 1966, although he was only referring to the last few years of Czechoslovakia. He praised the exceptions: "The walls of isolation were breached. There were, after all, many points of contact. There was a 'social club' in one of the palais on Na příkopě which was open to both languages and subsidized by the government. Also, German speakers went to the Czech theatre and concerts and vice versa. It was perfectly natural for some German papers (not all) to cover fully all the events in Czech cultural life (theatre, music, art, literature) – and vice versa too." Brod believed that before 1938–39 he could see on the horizon the possibility of the two cultures converging and working together in a European context, in particular when **Max Reinhardt** produced **František Langer's** *Periphery* in Berlin, but in the end he had to admit: "Unfortunately the possibilities were only hinted at. The point where the parallel lines would meet was never reached."

At the end of the 1940s, when Czechoslovakia became a socialist country, literary life was monopolised by the demands for social realism in literature. Publication was concentrated in relatively few publishing houses and journals, and the influential Czechoslovak Writers' Association was formed. Now literary activity in the city was concentrated in the Národní třída, where the Writers' Association was to be found in no.11, and in no. 9 the main publisher for literature written in Czech, the *Čecskoslovensky spisovatel*, together with **Odeon**, the main publisher foreign literature, and **Albatros**, the main publisher of books for children and young people.

Where Albatros is now, used to be the site of the Café Union, in which writers and artists of almost every field met – but it no longer exists. You can find Albatros quite easily by following the neon signs in the Národni třída: NEJLEPŠí DETEM (the best for children). This refers to the self-imposed duty of the authors, illustrators and editors to provide children and young people with nothing but the best books.

Josef Lada, whose illustration of *Schwejk* is recognised throughout the world, and **Jiří Trnka**, whose pastel drawings with their childlike appeal became a successful export, used to be familiar faces at Albatros. Now they have been replaced by **Ota Hofman**, whose *Pan Tau* has charmed children everywhere, and **Otakar Chaloupka**, whose ideas on literature for children and young people have had considerable influence.

Visitors to Prague who are looking for personal traces of and contact with literary figures are very likely to meet one or other of the key creative figures in the Czech literary scene somewhere in the Národni třída. Some of them can be immediately recognised as such just by their appearance. The well-dressed grey hair stands up slightly from the scalp, the clothing is both neat and casual, the expression soft and the posture strangely upright and bowed at the same time. They hurry to their appointments – in the editorial offices of the publishing house or in the restaurant of the "club" of the Writers' Association. In the years before the war, this "Club" was the Café National (*Národní kavárna*). In the years just after the war the "Club" still had guests such as Nezval, Halas and of course **Jaroslav Seifert**. Seifert continued writing until well into the 1980s and saw much of his later poetry translated into English and other languages. He was awarded the Nobel prize for literature in 1984. You could add name after name from the "Prague Spring" of 1968. Some of them lost their entitlement to be members of the "Club", and some live or publish abroad.

Left, in the baroque hall of the library in the Clementinum. Above, Egon Erwin Kisch, the "roving reporter".

However, most have stayed, in the home country of their language, in their city of Prague. Sometimes they drink coffee in the Café Slavia, a few yards on towards the Vltava, next to the National Theatre, where Jaroslav Seifert found inspiration for the "Café Slavia" poems:

Through the secret door from the Vltava quay,
which was of such transparent glass
that it was almost invisible,
and whose hinges
were smeared with oil of roses,
Guillaume Apollinaire would sometimes come.
His head was still bandaged,
from the war.
He sat down with us
and read beautiful, brutal poems,
which Karel Teige translated on the spot.
In honour of the poet
we drank absinthe.
It was greener
than any other green,
and if we looked from our table
through the window,
the Seine flowed past the quayside.
Ah yes, the Seine!
And not far away, on broad-spread legs,
the Eiffel Tower stood.
Once Nezval came, wearing a stiff hat.
We didn't know it then
and he didn't know either,
that Apollinaire was wearing the same one
when once upon a time he fell in love
with the beautiful Louise de Coligny-Chatillon,
whom he called Lou.

The star among the writers of today, however, is **Bohumil Hrabal**, the literary creator of the "Bafler", whose *Uncle Pepin* has brought another Schwejk to life. The "star" Hrabal holds court, democratically, in the pub *U zlatého tygra* (The Golden Tiger), not far from the Národní třída.

CITY OF MUSIC

Bohemia has a rich musical past, and it has brought Prague the reputation of a "musical city". However, the city – the "musical heart of Europe", as David Oistrakh once said – doesn't just owe its reputation to the "big names" – which, after all, grace many other cities. Rather, the important position that is accorded to Prague by statements such as Oistrakh's is based on a long tradition of musical culture.

Bohemian musicians: The great flowering of Czech music, also known as "Bohemian classicism", took place in the 18th century. The saying "All Bohemians are musicians" dates from this time. The proverbial musicality of this nation is probably due to the fact that in Bohemia, support for musical education was widespread. Documents dating from this time show that most cantors (school-teachers) had a musical education and saw to it that nearly every pupil could play an instrument or at least sing.

The *General Musical Journal* of 1800 says: "A great number of these cantors were truly skilled and talented musicians... The vast number of skilled Bohemian musicians, of which one can find no better proof than by reading the lists of players in the court orchestras of Europe, is explained by the fact that the best of the nobility insist that all their servants – from estate manager down to stable boy – should enjoy music and be able to play an instrument properly."

Musical talent and education, in those early days, was an excellent means of obtaining material comforts and advantages. A position as a servant freed a peasant from serfdom and from military service, and those who proved themselves to be good musicians had the hope of being released from the service eventually.

When the English music writer Charles Burnley visited Bohemia in 1772 he was so surprised by the level of musical knowledge and skill in the country that he named it the "conservatory of Europe". The fruitful musical climate produced not only many folk musicians but also a surplus of trained performers, who had difficulty in earning their living in their home country.

The unstable political situation in the Bohemia of the 18th century and religious persecution forced many people to emigrate. Countless musicians also left the country, and, thanks to their skills, easily found work all over Europe. Everywhere these emigrants went, they commanded respect, influenced the new instrumental style of Classicism, and left definite traces in the structure of its melodies. On the other hand, Bohemian music became exposed to foreign influences, which it in turn incorporated.

Mozart in Prague: The visits of Wolfgang Amadeus Mozart, who found many friends in Prague, should also be seen in this context. Mozart tried to build up a secure career for himself in Vienna, but the Viennese public and the imperial court mainly reacted with incomprehension and indifference. At this time he received news from Prague about the reception of his opera *The Marriage of Figaro*. An invitation quickly followed, and he took it up in early 1787.

In Prague he witnessed Figaro-fever, which had gripped the whole city. Apart from this, the visit brought Mozart a contract for an opera. It was to be *Don Giovanni.* It was commissioned by the impresario of what was then the Nostitz Theatre (now the Estates Theatre), which, in contrast to all the other theatres in Central Europe, was not tied to a court, but was a relatively independent institution. The fact that in Prague opera had been available to the general public for a long time explains the interest of the broad mass of people. The premiere in the autumn of the same year was a great success.

Music and middle classes: From the early 19th century on, the aristocracy of Prague gradually lost their position as the most important patrons of the arts. The rising middle classes claimed their share in the process of shaping cultural life. The centre of activity

Preceding pages: all Bohemians are musicians. Left, the composer Bedřich Smetana.

moved from aristocratic salons to public concert halls, and a new era dawned. It was formed by two institutions which left a definitive mark. One was the Society of Artists, founded in 1803 and modelled on its predecessor in Vienna, the other was the Prague Conservatory, which opened in 1811, was the first in Central Europe and set the standards for the rest. The city, which was still under the strong influence of the Mozart cult, was exposed to new influences. Carl Maria von Weber, who was director of the Nostitz Theatre from 1813 to 1816, acquainted Prague with Beethoven's *Fidelio* and the first Romantic operas. In the same house, Niccolo Paganini celebrated great successes. Concerts also took place in the Konvikt, a complex in the Bartoloméjská ul. which is being restored. Beethoven appeared here. Later, a concert hall on the Slavic island (*Slovanský ostrov*) became a venue for Hector Berlioz, Richard Wagner and Franz Liszt. Liszt also played in the Platyz *(Uhelný trh 11)*. The scene was set by the mighty flood of largely German music, and Czech music faded into the background.

Smetana and Dvořák: The awakening of Czech national consciousness during the politically and economically troubled times of the early 19th century saw the first generation of Czech artists faced with the task of creating their own culture, which did not establish itself until the second half of the century.

The name of Bedřich Smetana (1824–84) is inextricably bound up with Prague, and in his work Czech music reached its first peak. Born in Litomyšl, Smetana came to Prague to study music. During the Czech nationalist rebellion in 1848, in which he personally took part, his patriotic feelings awoke. His wish was to unite the highest artistic expectations with the demands of an independent national culture. However, the way to this goal, which he was to achieve most of all in his operas, was long and difficult. Apart from a five-year stay in Göteborg in Sweden, Smetana took part wholeheartedly in the musical life of Prague, but at first he tried in vain to establish himself as a conductor and composer. Not until the success of his opera *The Bartered Bride* did he achieve the desirable position of conductor to the Czech Opera and widespread recognition, which, however, did not remain uncontested. After the loss of his hearing Smetana had to give up his career as a practising musician, but continued to compose and created some notable works, including the series of symphonic poems entitled *Má Vlast* (My Fatherland) and the string quartet *Aus meinem Leben* (From My Life). Bedřich Smetana received the highest honour when his opera *Libuše* was performed at the official opening of the National Theatre, a ceremony that symbolised the peak of national aspirations.

While everyone in Prague was raving about Smetana, another Czech composer had already started to show his talent. Antonín Dvořák (1841–1904) was born near Prague and first attracted attention with his *Hymnus*, a nationalistic cantata based on Halek's poem *The Heroes of the White Mountain*. He attended the organ school in Prague and was organist at St Adalbert's from 1874 to 1877, during which time he wrote a number of compositions. He was recognised by Brahms, who introduced his music to Vienna, sponsoring the publication of the *Klänge aus Mähren* (Sounds from Moravia), which were followed by a commissioned work, *Slavonic Dances*. His *Stabat Mater*, performed for the first time in London in 1883, won him European acclaim. In 1891 he was appointed director of the New York conservatory. His ninth symphony *From the New World* retained a distinct Slavonic flavour. Returning to Prague in 1895, Dvořák remained true to the musical traditions of Bohemia until his death and had great influence on the musical life of the city.

Left, bust of Antonín Dvořák. Above, the former Nostitz Theatre (now the Estates Theatre).

Modern music: However, the strong flow of national culture did not have a detrimental effect on Prague's open-minded attitude to modern European music. Gustav Mahler, in 1885 conductor of the New German Theatre's orchestra (today it's the Smetana Theatre), had the first performance of his 7th Symphony take place in Prague. The same theatre was directed from 1911 to 1927 by Alexander von Zemlinsky, who had close contacts with centres of music in Vienna and Berlin and acted as a go-between. By this means Alban Berg and Arnold Schoenberg, among others, had the opportunity to get to know the Prague music scene. One result of these busy cultural exchanges was the premiere of Schoenberg's *Ewartung* (Expectation), which took place in Prague.

Concerts In Prague

If you want to hear classical music in Prague, you won't be disappointed. A varied programme is assured by several symphony orchestras, a number of chamber music ensembles, two opera houses and many soloists, together with visiting foreign musicians. Up-to-date information is provided by the Prague Information Service (Na příkopě 20) and posters put up all over the city.

The traditional repertoire is dominant in the programmes of the (always well-filled) concert halls, and Czech composers, both old and new, feature prominently. Opera performances are somewhat overshadowed by orchestral music, which everywhere lives up to the highest expectations. For those who like chamber music, the various string quartets, among other groups, can be strongly recommended. They are remarkable for their musical excellence. Regular series of concerts of course also feature those ensembles based in Prague who have received international recognition. Among them, without any doubt, is the Czech Philharmonic.

Musical events always lead visitors to interesting places in Prague. In this way they are offered an excellent opportunity of experiencing many of the "sights" in a quite different way, of seeing the interiors and of spending time in places which are definitely worth seeing. Apart from the great concert halls and the opera houses, these are usually churches, palaces and palace gardens. For example, the effect of the bold architecture of St Vitus' Cathedral is enhanced by the addition of a large orchestra and choir, and the polyphonic music of medieval times breathes new life into the Romanesque basilica of St George. The bare, impersonal interior of the Bethlehem Chapel suddenly loses its museum-like character when it becomes a concert hall, and the baroque splendour of St James's Church in the Old Town only really unfolds during the organ concerts, which take place regularly every Tuesday afternoon.

Historic rooms that are normally inaccessible are opened for musical events, for instance those in the newly restored Palais Martinic in Hradčany Square, or the Great Hall in the Palais Waldstein, the mirrored chapel of the Clementinum and others. Music draws people to the *Slovanský ostrov* island, to the Riding School in the castle, to the St Agnes Convent complex, or to the Hvežda summer palace (Letohrádek Hvežda – on the western edge of town).

In summer, open-air concerts are added to the list, in the gardens of the Palais Waldstein, in the Maltese Gardens of the Museum of Musical Instruments in the Malá Strana, or in the palace gardens below the castle. In the baroque summer palace of the Villa Amerika, the Dvořák Museum, you can listen to music surrounded by mementos of the composer, as you can in the Villa Bertramka and its garden, where Mozart completed his operas *Don Giovanni* and *La Clemenza de Tito*. Mozart's venue in Prague, the old Nostitz Theatre (now the Estates Theatre) reopened in 1991 after extensive renovations. In the "Music Theatre – Lyra Pragensis" (Opletova ul. 5), performances take place which take themes from a great variety of musical experiences and present them to the public in the form of film or sound recording.

Once a year the posters in Prague are dominated by a white "f" on a blue background. The symbol, which looks like the sound aperture of a violin, announces the "Prague Spring". The musical life of Prague culminates in this famous festival, which has a tradition going back over 40 years and a fixed place in the international calendar of festivals. On 12 May events mark the anniversary of the death of Bedřich Smetana (1824–1884), the founder of the Czech national music movement of the 19th century. This cultural occasion is regularly opened with a performance of the composer's *Ma Vlast* (My Fatherland), a cycle of symphonic poems, composed between 1874 and 1879. The best known of the pieces is the second poem, *Vltava*, which describes the course of the Vltava river.

Mozart im Hause Du-
schek,

PUMA
DHL

TENNIS IN CZECHOSLOVAKIA

Following its successes over the past 25 years, the reputation of the Czechoslovak tennis style is recognised throughout the world. But only experts and true tennis fans know that the roots of Czechoslovak tennis lie much deeper and date back to the 19th century; they share the same age and the same traditions as, for example, those in Britain.

The oldest club is the ČLTK, founded in 1893, which was a member of the British Lawn Tennis Association from 1894 to 1906. The most famous players of earlier times were K. Koželuh, professional world champion in 1929, 1932 and 1937, and J. Drobný, who won the Wimbledon championship in 1954 and reached the final in 1952 and 1949. V. Suková, the mother of Helena Suková, also reached the ladies' final in Wimbledon, and Jan Kodeš won the men's singles. It is impossible to imagine modern tennis without names such as Martina Navrátilová, Ivan Lendl or Miloslav Mečír.

After Jan Kodeš' Wimbledon victory in the men's final in 1973, tennis became very popular in Czechoslovakia. Young people were encouraged to play. Matches were organised under the heading "The Search for New Kodeš' and Sukovás". More than 15,000 young people aged between 9 and 15 took part.

Today the number of players organised in clubs has reached 60,000, and perhaps another 60,000 play for fun. There are 810 clubs with 3,500 places in the country.

Special training methods, long-term tuition of young players and the high standard of the Czechoslovak coaches have given the country an important position in today's tennis world. The Czechoslovakian Tennis School has an international reputation, and Steffi Graf probably wouldn't have got where she did were it not for the efforts of her trainer, Pavel Složil.

To explain the success of the Czechoslovak style of tennis, a reminder of a few facts about the most notable players may be necessary. **Martina Navrátilová** was born on 18 October 1956 in Prague and grew up in Revnice, a village not far from the city. Her first coach was her stepfather. She started her career as Junior Champion in 1972, and in 1973 she was in the junior final at Wimbledon. In 1975 she left Czechoslovakia and became an American citizen. Martina Navrátilová was the third woman to win the "Grand Slam" in Wimbledon eight times. In 1984, Navrátilová won 74 matches in succession and was defeated in the 75th by another Czechoslovak player, Helena Suková.

Ivan Lendl is proof that Prague is not the only source of good players. He was born on 7 March 1960 in Ostrava in northern Moravia, and in the 1970s he was the most famous player in the country. He was awarded the title "World Junior Tennis Champion" in 1978. Lendl's talent for tennis is the work of his parents. His mother was a successful tennis player, who became national champion in the doubles in 1964 and 1969. Jimmy Connors once remarked that Lendl looked "like a half-boiled chicken". Lendl's answer was plain: he beat Connors at Flushing Meadow and had a half-cooked chicken delivered to Connors' room. Lendl has not yet achieved his great dream, a Wimbledon victory.

Jan Kodeš was born on 1 March, 1946 in Prague. In 1964 he was Junior Champion of Czechoslovakia, in 1973 he won in Wimbledon. In 1980 he and the country's team won the Davis Cup. For 11 years he was the no. 1 of Czechoslovak tennis. Today Kodeš, who has done a great deal for the game, is director of the new tennis stadium in the Štvanice district of Prague.

eceding ges: from e travel urnal of olfgang nadeus ozart. **ft**, a nner at nnis – artina vrátilová.

PUNKS IN PRAGUE

Yes, they do exist – punks in Prague. You may not find quite such a conspicuous youth scene as in many cities of Western Europe, but you do find traces of most youth cults. For years, music was a refuge for most of the young people, not only in Prague but throughout the country. They cared little for political protests or ideological debates. Forced to participate in official youth organisations, young people withrew by turning to their own interests, such as music. Sceptical of the world around them, they retreated into their own private world. Only seldom did optimism show its face, as in the lyrics of a song by the group *Laura and the Tigers*: "We are the new generation, bound by conventions and rules. We look for heaven and find earth, and life."

The overthrow of Stalinism had an electrifying effect on the music scene. Suddenly, the young people stood at the forefront of the demonstrations; it was as if the scene had awakened from a long hibernation. Today, you can still see a wide selection of Western music in the record shops, but the number of local bands is on the rise, and the music one hears in the discos is straight from the latest national charts. In the small cultural centres of individual districts of the city, concerts are being given wide range, from Heavy Metal music to avant-garde.

Even the cream of the international rock business is once again thinking of Prague. One of the first to come was Joan Baez; the Rolling Stones and Frank Zappa even had an official audience with the president. It didn't take long for Prague to be included in the standard programme of European tours. The rock musician and occasional member of parliament Michal Kocáb has seen to that.

Discos in Prague: Prague does have a disco and nightlife scene – that soon becomes evident to anyone visiting Wenceslas Square. The latest pop videos from abroad are shown in the video discos, and, just like in the Golden West, a bouncer is necessary to sort out those who can come in from those who can't. However, you shouldn't expect too much from a video disco. Often there are only one or two TV sets fixed above the dance floor. "No jeans or trainers" say the signs on the door of the **Discotheque Zlatá Husa**. If you're prepared to splash out, you can dance to rock music till the small hours in the **Jalta Club**, in the **Hotel Jalta**, also in Wenceslas Square. However, the place is really just a meeting point for Western tourists. Except for the few young ladies who offer to escort just about every Westener that goes through the door, this isn't the place to come if you want to meet the "normal" youth of Prague: the pleasure is quite unaffordable for locals on average earnings.

The **Video Disco Alfa** on the opposite side of Wenceslas Square is better for meeting people, or you can go to the **Tatran**, which has a plain glass floor lit with coloured lights. Here you still have to pay a few crowns at the door, but the atmosphere is much more relaxed and friendly than in the Zlatá Husa or the Jalta. The discos in the **Botels** *(Nábř. L. Svobody)* are also very popular with young people in Prague. In the **Admiral** and the **Albatros**, for instance, you can dance the night away without having to spend too much.

If you really want to get away from tourists for once, try the **U Holubů**, Prague 5, S.M. Kirova. For a few crowns you can really let your hair down and dance till two in the morning. If your taste runs to punk or New Wave, take yourself off to the **Na Chmelnici**, Prague 3, 10 Koněva. Some members of this scene meet at the monument to St Wenceslas.

Right, you can hear Heavy Metal sounds in Prague too

In summer, the Charles Bridge is another favourite meeting place. Young people meet here to play the guitar together and have a short jam session. There's no rigid division in Prague between the various "scenes". There are no pubs exclusively for teenagers or for punks. You meet your friends and go to any old pub. If this happens to be renovated or converted to a restaurant, you go somewhere else. In the outlying districts of Prague there are smaller discos or places where young people meet, which advertise with posters on the billboards around construction sites.

Heavy Metal fans who feel compelled to listen to the Czech Hard Rock groups *Citron* or *Vitacit* should go every Thursday to the **House of Culture**, *Kulturní dům Barikádniků*, Saratovská 1, Prague 10. Most of the concerts of this sort take place here, and there's also a Heavy Metal disco. Beer and Becherovka flow in abundance, and sometimes the place can be a real buzz. Of course, concerts don't always have to be loud and heavy. In Prague young rock groups such as *Stromboli*, *ETC* or *Vyběr* are also very popular. However, there is no one place in which to hear these groups. Your best bet is to ask in the **Sluna** box offices in the Alfa Arcade or in the Lucerna Arcade in Wenceslas Square. Here you can get tickets for all sorts of events: rock, jazz, chamber orchestra or song recitals. Occasionally small rock concerts take place in the **Lucerna Palace**.

Folk and jazz: Concerts on a larger scale take place in the **Palace of Culture**, *Palác Kultury*, or the **Sports Hall** at the exhibition grounds (*Výstaviště*, Prague 7, Metro station *Vltavská*). In summer free open-air festivals lasting two days are held here. In April a rock festival lasting several days is held in the Palace of Culture. Folkrock and blues concerts are also very popular.

After a pop concert in the Prague House of Culture.

Country, folk and traditional folk music concerts are held in the **Sophia Hall** on the Vltava island of Slovanský ostrov.

With a bit of luck, the very best jazz can also be heard in Prague. From time to time the Prague Jazz Days are held, and musicians of international repute come to perform. However, the local jazz scene also has plenty to offer.

When Milan Svobada invites his guests to the **Malostranská beseda** (the Malá Strana House of Culture), the occasion is always a huge success. This place, right in the Malá Strana Square, is one of the best addresses for good jazz in the whole of Prague. Apart from Milan Svoboda, Martin Kratochvíl and Jana Koubkova regularly appear here. There is also some sort of event every two or three days in the Jazzclub.

The second best address for jazz in Prague is bound to be the **Club Reduta** in the Národní třída, the National Street. A somewhat lower quality is on offer in the cellar bar on the other side of the street, the **Metro Jazz Club**. Large-scale events such as the "Jazzparanto" with Jana Koubkova take place in the Palace of Culture. The jazz pub **Prazan**, in the exhibition grounds, also puts on a variety of jazz events.

Who's who: "New Romantics" and Prague's young professionals meet in the Grill Room of the **ČKD Dům**, right by the exit from the Metro station *Můstek* in Wenceslas Square. It's *de rigueur* to be overdressed in the best designer clothes – just like their counterparts in the West.

If you like Art Nouveau and prefer the company of young men, you won't go far wrong in the **Café Evropa**. This coffee house is open from 6am to midnight and isn't only a place for making this sort of contact. Just as in the other great coffee house in Prague, the Café Slavia, the clientele is very colourful and mixed.

**ler
ting is
nmer
ning for
hockey
yers.**

12
1021
servis pletacích strojů
z dovozu
domácí potřeby Praha
středisko služeb

The numbers in brackets refer to the map in the Travel Tips section, page 230.

At the beginning of the 20th century in Europe, modern styles of architecture were evolving. Prague was, along with Paris, Vienna and Berlin, one of the places where the artistic avant-garde were using revolutionary ideas and manifestos to shape architectural styles so that they reflected their ideals of the world. A progressive atmosphere, such as one can hardly imagine today, was created by close economic and cultural ties between the cities.

Architects from Germany (Peter Behrens) and Vienna (Adolf Loos) designed important buildings for the city. Le Corbusier, Walter Gropius and Hannes Mayer held influential lectures here, and Czech architecture soon won itself a place in European artistic journals. In this environment an extremely creative, internationally influential and yet Bohemian-inspired architecture could develop to mark an important and fascinating chapter in the history of Prague's development. It is sometimes more exciting and challenging to discover the monuments to this overturning of the artistic establishment than to visit the monuments of long-vanished ages.

As in Berlin and Vienna, modern architecture in Prague developed out of the historicism left over from the 19th century, which gives whole districts of Prague their characteristic appearance. The first liberation from this style, and the beginning of the modern period, was the advent of Art Nouveau with its natural style of ornament, instead of the over-used curlicues of historicism. The most beautiful example is considered to be the **Hotel Evropa** ***(1)***, built in 1900.

The man who did most to initiate the U-turn in architecture was the Viennese master architect Otto Wagner, whose student Jan Kotěra was one of the first to introduce new trends into Prague. At first he was strongly influenced by Vienna, as can be seen in the **Peterka House** ***(2)*** at the end of Wenceslas Square, which he built in only one year in 1900. With its tall slender windows and elegant ornamentation, it is considered to be one of the finest and most classical examples of Art Nouveau architecture in the city. He later combined new ideas with Bohemian characteristics, creating a new architectural language, as in the **Urbánek House** ***(3)***, built in 1912, where ornamentation has retreated into the background in favour of the effect of the material, bricks and copper.

The Stenc House ***(4)*** by the Kotěra pupil Otakar Novotný, with the finest of brick facades built with exquisite restraint, dates from the same time. Here, architecture is not understood as a matter of ornament, but as a play of proportions, light and shade.

The Wagner pupil Hubschmann marked the courageous transition to a plainer architecture more suited to the new times with his **apartment house** by the Jewish Cemetery ***(5)***, built in 1911. Further development of Kotěra's ideas can be seen in the **office block** ***(6)*** built in 1924, in which Kotěra consciously incorporates baroque forms into his building.

Cubism and Rondo-Cubism: The second chapter of modern architecture in Prague began with the exhibition of the first Cubist paintings by Picasso and Braque in Paris. These changed not only the development of painting, but also the style of buildings in Prague. In the revolutionary atmosphere of the times, people saw Cubism, with its faceted and prism-like dissection and abstraction of surfaces, as a possible way of overcoming old conventions and projecting an image of the new age.

Josef Chochol became a notable architect of the times with his buildings in the **Neklanova 2** and **34**. Another example from the inner city is the **street lamp with seat** ***(7)*** constructed in 1913 by Vladislav Hofmann, which is delightfully linked with the church of St Mary of the Snows. An early example

Preceding pages: house facades in the Pařížská. Left, Cubist elements.

of the period dates from the year 1912. It is the **House of the Black Madonna *(8)***, by Josef Gočár. The building lends the little square a distinctive note and demonstrates that even in those days it was possible to build anew in the modern idiom without destroying the optical harmony of an entire district. By the way, there is a lovely café on the first floor.

The Cubist influence came at a time when the country was in the process of dissolving its ties to the Habsburg monarchy and founding the Czechoslovak state. This was the political background to the efforts of leading architects, who were trying to create an independent national architecture using Cubist methods combined with forms of vernacular architecture such as arches, cylinders and shapes in high relief.

This style is known to art historians as Rondo-Cubism and has had a strong influence on many buildings in Prague. Notable examples of this style are the **office block *(9)*** by Pavel Janák, dating from 1922, and its neighbour dating from 1923 ***(10)***, shortened to give a monumental perspective and by the same architect. Both of these buildings are situated opposite a house by Kotěra which was built some 10 years earlier. The master architect, Novotný, also designed numerous buildings in this particular style, for example, the **hostel *(11)*** with its new-style, and very effective, pleasant colour scheme.

Modern architecture: During the course of the 1920s, European architecture took another direction, and the white "classic" style of modern architecture prevailed over the nationalist rondo-cubist style. The third chapter of modern architecture in Prague began. A symbol of this new direction, and of the cosmopolitan character of Prague at that time, is the light, almost disembodied appearance of the **Mánes House of Artists *(12)*** dating from 1928. This radiant, white, puristically plain building was confidently designed by Novotný to protrude into the Vltava.

An interesting succession of interiors dating from this third period is provided by the shopping arcade **Black Rose *(13)*** by Oldřich Tyl, dating from 1929. Glass tiles were used for the first time in its roof. The galleries can be reached by a double spiral of staircases. Unfortunately, the arcade has lost some of its welcoming appearance due to lack of maintenance.

Towards the end of the 1920s the modern architectural style of steel, glass and smooth surfaces achieved widespread recognition in Prague as it did in other places, even if not all these "modern" buildings are still admired today. Notable precursors of this pre-Munich period are the **Hotel Juliš *(14)*** dating from 1933 and the **Lindt House *(15)*** dating from 1927.

Post-war architecture: The architecture of the post-war years created few buildings that are still admired today. One exception is the **MAJ Department Store *(16)***, designed in 1968 by the architects' collective *Stavoprojekt Liberec*. With the generous dimensions of its escalator and its acknowledgement of modern building materials, the building has attracted much admiration and respect both at home and abroad.

Left, apartment house in the Neclanova. Right, escalator in the MAJ department store.

Plzeňský Prazdroj
PLZEŇ

PLZEŇS

AROUND PRAGUE

There are a number of interesting places around Prague that are well worth a visit. Many of them are close enough for a day trip. One of the most famous destinations is **Karlštejn Castle**, about 21 miles (35 km) from Prague in the direction of Plzeň. Karlštejn was built by Charles IV as the representative residence of the king and the depository of the Bohemian crown jewels.

The foundation stone was laid on 10 June 1348 and the mighty castle, considered to be one of the most beautiful in Europe, was finished within 10 years. Of the various halls and rooms that are open to the public nowadays, the **Chapel of the Holy Cross** is perhaps the most remarkable. Its walls are richly decorated with 2,450 precious and semi-precious stones.

Another castle, **Castle Křivoklát**, lies in the direction of the E 12 highway about 21 miles (35 km) from Prague. Originally this was a little wooden hunting lodge. Charles IV had this extended into a castle especially for his wife Blanche. Nowadays the castle contains a museum with musical instruments and a gallery of paintings. In the nearby village of **Lány** is the grave of the founder of the Czechoslovak republic, Tomáš Masaryk. The countryside around Karlštejn and Křivoklát is excellent for walks.

The palace of **Konopiště**, near the town of Benešov, is famous for its extensive collection of weapons and hunting trophies. **Mělník**, some 19 miles (32 km) away from Prague, lies at the confluence of the Vltava and the Elbe (Labe). The area is particularly famous for its vineyards.

The park of Průhonice, on the southeastern edge of Prague, is one of the

Preceding pages: transportin beer. Belo Mělník.

largest and most beautiful parks in the whole of Europe. It covers an area of 494 acres (200 hectares) and its varied landscape contains 7,000 rhododendrons, azaleas and other shrubs. It is particularly well worth visiting in May and June, when the shrubs blossom.

A little out of the way, in Prague 7 (opposite the Zoological Gardens), lies the palace of Troja, a summer retreat built for Count Sternberg. It is famous for its opulent interior decoration and its beautiful staircase with scenes of the battle between gods and Titans.

Also on the outskirts of Prague, in Prague 6, is the **Star Palace** (*Hvězda*), whose name derives from its star-shaped ground plan. The palace, which originally lay in the royal hunting grounds, was built in 1555–56. In the 18th century it served as a gunpowder magazine. It was restored in 1949–51 and is now a museum for the works of Mikoláš Aleš and Alois Jirásek.

On the road to Kladno, some 16 miles (25 km) from Prague, is the village of **Lidice**, which was burned to the ground in 1942 after the assassination of the "Reich Protector" Reinhard Heydrich. Today there is a memorial to all those who died in the accompanying massacre, and a museum.

About 44 miles (70 km) from Prague in an easterly direction lies the very remarkable town of **Kutná Hora**, which once rivalled Prague in its development of civic splendour. Some of the buildings dating from this time have been preserved, among them the Foreign Court, *Vlašský dvůr*, which contains the royal mint. Also worthy of attention are the Gothic St Barbara's Church and the Stone House, *Kammený dům*, which is considered a masterpiece of medieval stonemason's art. If you're travelling to Prague via **Cheb** you should of course make a stop in the famous spa **Karlovy Vary** (Carlsbad).

llic untryside t outside ague.

TRAVEL TIPS

GETTING THERE

BY AIR

It is possible to book a flight from the following cities (amongst others) to Prague: Zurich, Geneva, Vienna, London, New York, Montreal and Toronto. Prague Ruzyně airport lies 20 km (13 miles) northwest of the city.

There are public bus services to the centre, as well as taxi and hire cars.

The state travel agency Čedok runs a shuttle bus service between the airport and the city's Interhotels between 11am and 4pm.

The national airline (ČSA) also operates a similar service between the airport and the city terminal (Vltava Travel Agency, Revolucni 25. Tel: 231 7395, 2146). It runs every 30 minutes, Monday–Friday 5.30am–6.30pm, Saturday and Sunday 6.30am–6.30pm. The journey takes 30 minutes and costs 15 crowns (Kčs). The buses also stop at the terminal station *Dejvická* of the green Metro line A.

AIRLINES

ČSA – Československé Aerolinie
Tickets and reservations: Revoluční 1 (*Kotva*) Prague 1. Tel: 2146 (reservations); 232 2006 (domestic tickets). Flight information: Revoluční 25 (*Vltava*) Prague 1. Tel: 231 7395, 2146. Both offices are located near the Metro station *Náměstí Republiky*.

Central information service, Ruzyně Airport, Prague 6. Tel: 367 760, 367 814, 334 1111.

Aeroflot, Na příkopě 20, Prague 1. Tel: 232 4707, 260 862. Airport: Tel: 367 815.
Air Algerie, Žitná 23, Prague 1. Tel: 265 483, 275 770. Airport: Tel: 225 770.
Air France, Václavské náměstí 10, Prague 1. Tel: 260 155. Airport: Tel: 367 819.
Air India, Václavské náměstí 15, Prague 1. Tel: 223 854.
Alitalia, Revoluční 5, Prague 1. Tel: 231 0535.
Austrian Airlines, Prague 1, Tel: 231 2795. Airport: Tel: 231 6469, 367 818.
British Airways, Stěpánská 63, Prague 1. Tel: 236 0353.
Delta, Pařížska 11. Tel: 232 4772.
KLM, Václavské náměstí 39, Prague 1. Tel: 264 362, 264 369. Airport: Tel: 367 822.
Lufthansa, Pařížska 28, Prague 1. Tel: 231 7440, 231 7551. Airport: Tel: 367 827.
SAS, Stěpánská 61, Prague 1. Tel: 228 141. Airport: Tel: 367 817.
Swissair, Pařížska 11, Prague 1. Tel: 232 4707. Airport: Tel: 367 809.

CSA OFFICES ABROAD

Austria: Parkring 12, 1010 Vienna. Tel: 01-523 805 or 512 9886.
Canada: 2020 rue Universite, Montreal, Quebec H3A 2A5. Tel: 514-844 4200 or 844 6376; 401 Bay Street, Suite 1510, Toronto, Ontario M5H2Y4. Tel: 416-363 3174 or 363 3516.
Germany: Baselerstrasse 46-48, 6000 Frankfurt/Main. Tel: 069-253 559.
Switzerland: Sumatrastrasse 25,8006 Zurich. Tel: 01-363 8000 to 363 8009; CSA Office 334, Postfach 219, 1215; Geneva Airport: Tel: 022-798 3330.
United Kingdom:12a Margaret Street, London W1N 7 LF. Tel: 071-255 1898 or 255 1366.
United States: 545 Fifth Avenue, New York, New York 10017. Tel: 212-682 7541 or 682 5833.

BY RAIL

There are direct train connections to Prague from Germany and Austria. From Stuttgart and Munich, the journey takes approximately 8 hours, from Frankfurt 10 hours, Berlin 6 hours, Hamburg 14 hours and Vienna 6 hours. All trains from Southern Germany and Austria come in at the Main Station (*Hlavní nádraží*). Trains from the direction of Berlin come to halt at the Masaryk Station (*Masarykovo nádraží*) or at Prague-Holešovice Station.

Other rail destinations in the country can be reached via Prague. Travellers who do not have a ticket from the capital city to their ultimate destination may purchase one at the main railway station in Prague.

Domestic and international tickets can be purchased in Western currency at:
Čedok, Na příkopě 18, Prague 1, or directly at the railway station in Czechoslovakian crowns.

Further information pertaining to rail travel can be obtained in Prague from 7am–3.30pm, tel: 236 4441; and 6am–10pm at the main station, tel: 235 3836; also Prague Smichov, Tel: 2161 5086.

THE MAIN STATION

Prague main station is clearly laid out on two floors. The lower level contains the counters for domestic tickets as well as the PIS information office and shops. International tickets are purchased in the upper hall, which also has the room booking agency and exchange bureau, AVE.

A large digital display board shows departures and arrivals. Toilets, showers and left luggage are located under the main hall, together with the luggage lockers which cost 4 Kčs.

Taxis line up outside the southern side exit.

BY ROAD

Travellers arriving from Germany, can reach Prague via the following main border crossings from:

Baryreuth via Schirnding/Pomezí (175 km/110 miles)

Nuremberg via Waidhaus/Rozvadov (171 km)

Regensburg via Furth im Wald/Folmava (172 km)

Passau via Phillipsreuth/Strážný (164 km/100 miles)

Munich via Bayrisch Eisenstein/Železná Rudá (171 km)

Berlin via Zinnwald/Cínovecb (90 km/55 miles)

The border can also be crossed at the following points (from north to south):

Bad Schandau/Hřenskol, Bahratal/Petrovice, Reitzenhain/Hora Sv. Šebastiána, Oberwiesenthal/Boží Dar, Bad Brambach/Vojtanov, Selb/Aš, Waldsassen/Svatý Kříž, Waldmünchen/Lísková, Eschklam/Všeruby and Haidmühle/Stožec (only for pedestrians and cyclists).

If you're entering Czechoslovakia from Austria (from west to east):

Salzburg via Linz Summerau/Horni Dvořistě (186 km/115 miles)

Vienna via Gmünd/České Velenice (195 km/120 miles) or Grametten/Nová Bystřice (177 km).

Further crossings are to be found at:

Weigetschlag/Studánky, Wollowitz/Dolni Dvořistě, Neu-Nagelberg/Halámky, Kleinhaugsdorf/Hatě, Laa a. d. Thaya/Hevlín, Drasenhofen/Mikulov, Hainburg/Bratislava, Berg/Petržalka.

Although there are nowhere near as many private cars as there are in Western Europe, the driver on his way to Prague still has to reckon with delays. The main roads are generally in good condition, but the many lorries using them can make progress very slow, especially by day.

All vehicle drivers are required to be in possession of a valid national driver's license, car registration documents and a car nationality sticker. The international green insurance card should also be taken along. At the border "citizens of foreign countries" are handed a special vehicle license which is to be filled out and then shown with the other documents. If the driver is not using his own vehicle, then he will have to provide written consent from the vehicle owner. Caravans, trailers and boats require no special customs documents. Controls at the border crossings are often very thorough, so be prepared for long waits, particularly during the high season.

Nowadays, unleaded petrol (95 octane) is obtainable at most larger filling stations. Note: neither petrol nor diesel bought in the country may be taken out in reserve cannisters. If you're travelling by night you should make sure you have enough petrol, as it may be impossible to find a filling station open.

By and large, the international traffic regulations apply here.

The motorways are toll-free.

Maximum speed limit within city boundaries is 60 kph (about 35 mph). On expressways it is 110 kph (about 65 mph), and on country roads 90 kph (roughly 55 mph).

If you get caught exceeding the speed limit you can count on paying a fine of about 500 Kčs.

Driving while under the influence of alcohol is absolutely prohibited in Czechoslovakia.

Children under 12 years of age are not allowed to sit in the front seat.

Seat belts must be fastened outside the built-up areas.

BREAKDOWN SERVICES

Although there is an organised breakdown service including over 31 emergency aid vehicles in the country, it is often difficult to obtain the necessary spare parts for foreign vehicles. Therefore, it's a good idea to purchase International Travel Cover from your own automobile association prior to your intended journey.

The headquarters of the Czech Breakdown Service can be contacted any time of the day or night in Prague at Limuzská 12a, tel: (02) 154 or 773 4553. The service is free to those with the necessary insurance cover.

The agency **Pragis Assistance**, a non-stop emergency for tourists and motorised visitors can be reached by calling (02) 758 115. Such services normally have people who speak English or German or at least can tell you the way to the next garage.

BY BUS

There is a large choice of organised coach tours from Germany, Austria and Italy. Such journeys generally offer two nights' accommodation as well as a tour programme. Czech buses regularly ply the routes from Frankfurt and Nuremberg, Munich and Vienna. There is an operator that travels from Frankfurt to Prague several times a week:

Deutsche Touring, Am Römerhof 17, 6000 Frankfurt 1. Tel: 069-79 03 248.

If you are in Munich, a Czech operator runs a service to Prague every Saturday (returns Fridays). A single journey will cost you DM 57; tickets can be obtained from: **Autobus Oberbayern**, Lenbachplatz 1, 8000 München 2. Tel: 089-55 80 61.

From Vienna, there is a bus service running on Sundays and Wednesdays to the health resort of Piest'any. It costs 215 Austrian Schillings; further information can be obtained from Čedok.

The terminus is the Prague-Florenc bus station. Information regarding national connections can be obtained daily 6am–8pm by calling this station on 221445-9 (only Czech spoken). More gifted in languages are Autoturist (Tel: 290 956, 295 096, 204 300) and Bohemia Tours (Tel: 232 3989).

BY BOAT

The Cologne-Düsseldorf passenger shipping line offers cruises from Berlin and Hamburg via Dresden to Prague. These last between 4 and 7 days. The *Princess of Prussia* also plies the river Elbe, the journey taking a day from Dresden. In prague itself river cruises are offered on the Vltava. Further information may be obtained from the quayside under the Palacký Bridge: Prague 2, Nábrezí Rasinovo. Tel: 29 38 03, or 29 83 09.

There are no special formalities for those entering the country **by bicycle**, although it is advised to take out a good insurance.

TRAVEL ESSENTIALS

VISAS & PASSPORTS

For citizens of most European countries as well as the United States and Canada, no visa is required. Nationals of other countries are advised to contact their respective Czech embassies or consulates for information.

CUSTOMS

The Czech customs controls are quite rigid. In order to avoid misunderstandings, you should first inform yourself of anything you're unsure about beforehand. Upon entering the country, you'll be given a leaflet explaining the customs regulations.

IMPORT

All items of personal use may be taken in duty free; any electronic, photographic and filming equipment should be listed together with serial numbers and presented to the customs for confirmation. The list must be declared again upon departure. All items of personal use taken in to the country, must also be taken out.

You are allowed the following items for your own consumption (goods restricted to persons 18 years of age or older): 250 cigarettes, 50 cigars or 250 grams of tobacco, 2 litres of wine, 1 litre of spirits. Foreign visitors are permitted to take gifts into the country whose total value does not exceed 3,000 Kčs.

Hunters may take in 1,000 shot cartridges or 50 bullets for hunting purposes. Hunting rifles require a special weapons permit from the Czech authorities.

EXPORT

As of January 1992, the following items can be taken out of the country duty-free: 2 litres of wine, 1 litre of spirits, 250 cigarettes, 50 cigars or 250 grams of tobacco, items purchased in the republic the total value of which does not exceed 1,000 Kčs, articles that have been purchased in hard currency in TUZEX shops (make sure you save your receipts), cut glass, porcelain, souvenirs and other gifts when their value is not disproportionate to the amount of money you have officially changed and the reasonable cost of your staying in the country. Proof of such expenses can be provided by your hotel bill or an invoice from your travel agents.

In other cases, the export of goods can only go ahead with a customs permit, which costs between 3–150 percent of the price of the goods.

To export some items, it is not only necessary to have a permit but also to pay an additional fee. The list of these items, and of those whose export is forbidden, can be obtained at every border crossing. Antiques more than 50 years old can only be taken out with a special permit which can be very difficult to obtain. The best thing to do if you have any questions is to enquire at the customs office upon entering the country what exactly may be brought into the country and taken out, and whether these regulations may be subject to sudden change.

MONEY MATTERS

CURRENCY

The unit of currency is the crown (koruna or Kčs), which is divided into 100 haléř. There are 10, 20, 50, 100 and 1,000 crown notes and 1, 2, 5 and 100 crown coins as well as 5,10, 20 and 50 haléř coins. In August 1992 the exchange rate was approximately 6 Kčs to £1 sterling. There is no limit to the amount of hard currency that may be taken in or out of the country but crowns are not allowed to be taken in or out.

Eurochecks can be exchanged everywhere in the country and even on the border for a maximum of 6,500 Kčs. Sometimes what looks like an exceptionally good rate may be accompanied by an inordinately high commission. Travellers' Cheques are only accepted by banks and credit cards only by certain shops and hotels. It is possible to change back your Kčs, but in this case it is necessary to have an exchange receipt.

BLACK MARKET

While it is possible to exchange money on the "Black Market" just about anywhere in Prague, the practice is officially forbidden and offenders will be prosecuted. One should therefore not be tempted by any offers, particularly as one might also end up being cheated: the worthless Polish Zloty notes look conspicuously like crown notes.

EXCHANGE

In most shops and kiosks, payment is made in crowns. However, you may purchase items at the TUZEX shops in other currencies. Shops and hotels accepting credit cards will normally have the requisite signs on the door.

International exchange rates are published daily in the newspapers and displayed at banks and exchange bureaus. There is no shortage of places to change your money in the city centre. Banks are normally open from 9am–noon and 2pm–4pm.

FOREIGN BANKS

Société Générale, Paris, Prague 1, Vodičkova 34/1.

Creditanstalt-Bankverein, Vienna, Prague 2, Londynská 54.

Raiffeisen Zentralbank, Vienna, Prague 1, Gorkého nám. 3.

Länderbank AG, Vienna, Prague 1, Štupartská 646.

Vněšekonombank, Moskva, Prague 6, Na Zátorce 9.

Ljubljanska Banka, SFRJ, Prague 1, Lazarská 5.

Crédit Commerciale de France, Paris, Prague 1, Široká 9.

Dresdner Bank AG, Frankfurt, Prague 1, Panská 12 (Hotel Palace).

Deutsche Bank AG, Frankfurt, Prague 1, Národní třída 10.

Crédit Lyonnais, Paris, Prague 1, Nám. Curieových 43/5, (Hotel Intercontinental).

Banque Nationale de Paris, Prague 1, Na můstku 9, (ČKD-House).

Banque de l'Union Européenne, Paris, Prague 1, Vodičkova 17.

Scandinavian Banking Partners, Prague 1, Pařížská 25.

Citibank N. A., New York, Prague 6, Evropská 178.

Dialogbank, Moskva, Prague 1, Národní třída 10.

Financiere Credit Suisse, First Boston, Zug, Prague 1, V jámě 1.

Zentralsparkasse and **Kommerzialbank**, Vienna, Prague 1, Revoluční 15.

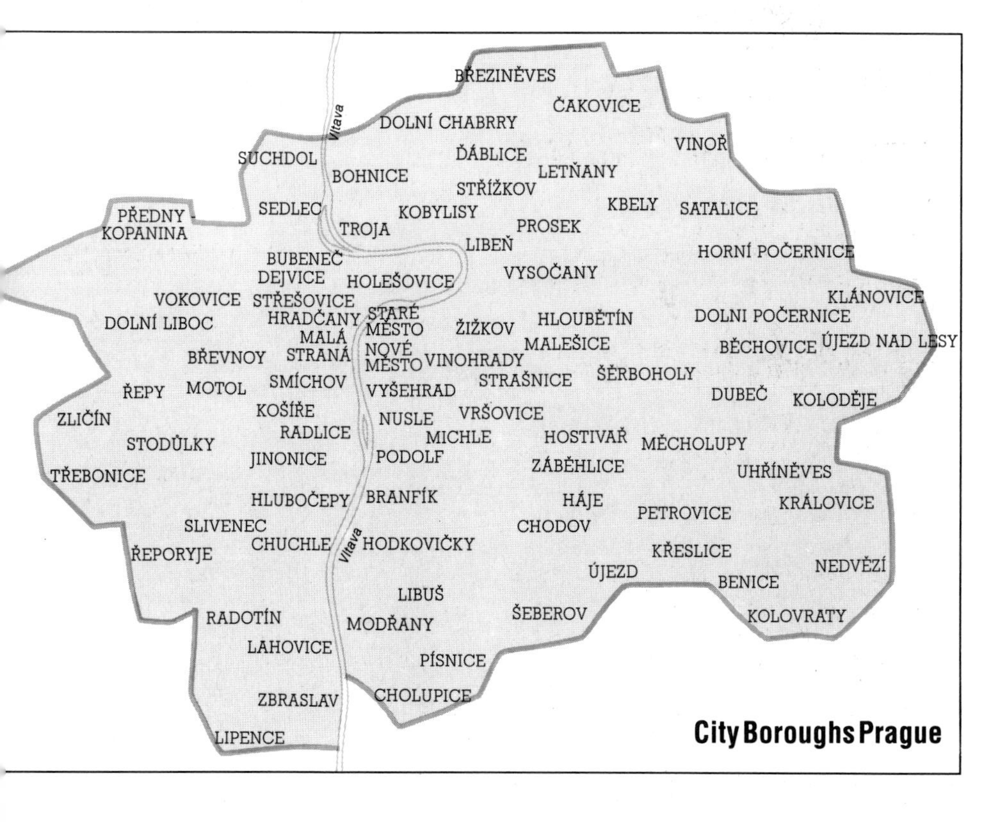

City Boroughs Prague

Getting Acquainted

GEOGRAPHY & POPULATION

The Bohemian metropolis is situated on the River Vltava (Vltava), spread out between seven hills. It lies between 176–397 metres (575–1,300 ft) above sea level, at 50° North and 14° East; about the same latitude as Frankfurt, Lands End and Vancouver. The city has a population of 1.2 million (1991) living over a total area of 497 sq. km (190 sq. miles). The historical part is made up of the Old Town (Staré Město), the New Town (Nové Město, Hradčany and the Lesser Quarter (Malá Strana), and boasts over 500 towers and steeples. The city's parks and gardens cover a total area of 870 hectares (2,150 acres). Prague is divided into 56 districts which are administered from town halls.

CLIMATE

Thanks to its protected position, the climate in Prague is particularly mild. The average annual temperature is 9.3°C (49°F). Summer: June 17.3°C (63°F), July 19.2°C (66°F); Winter: December 1.8°C (35°F), February 0.5°C (33°F). The average annual precipitation is 487 mm (19 inches) of which the least falls in February and the most in July.

ELECTRICITY

The electricity supply for the most part is AC 220 volts, although very occasionally there are outlets supplied with just 120 volts.

BUSINESS HOURS

Most **shops** are open weekdays 9am–7pm, with speciality stores being open from 10am–6pm. Smaller shops frequently close their doors for a couple of hours during lunchtime. On Saturdays shops close at noon or 1pm with the exception of large department stores, which often remain open until 6pm

Banks & Exchange Bureaux: In general most banks are open 8am–noon Monday–Friday. Larger branches, however, may stay open until 5pm.

Exchange bureaux are open 8am until at least 7pm; some even remain open until 10pm. Most hotels will exchange money around the clock, but be aware that their rates are slightly higher than at a regular exchange bureau or bank.

NATIONAL HOLIDAYS

The dates of national holidays at present are as follows:

1 January: New Year's Day
Easter Monday
1 May: May Day
5 July: Feast Day of SS Cyril and Methodius
6 July: Anniversary of the death of Jan Hus
28 October: Day of the Republic
25–26 December: Christmas

Various Christian holidays, for example the Feast of Corpus Christi and The Assumption of the Virgin Mary are celebrated in different regions but are not considered national holidays.

WHEN TO GO

From a climatic point of view, the best times to visit Prague are the spring and autumn. In May, when the gardens and parks are in full bloom, begins the classical music festival "Prague Spring". The mild autumn with its stable weather offers the best prospects for extended strolls around town.

Communications

POSTAL SERVICES

The General Post Office in Prague is open 24 hours a day. It is situated in a street off Wenceslas Square, half way along on the left (when viewed from the bottom): Hlavní pošta, Prague 1, Jindřisská 14. The other post offices are open 8am–7pm Monday–Friday and 8am–noon Saturday; smaller branches are generally only open 8am–1 or 3pm at the latest Monday–Friday. Prague is divided into 10 postal districts.

POSTAL CHARGES

Stamps can be bought in the post offices or at newspaper kiosks. Enquire about current postal rates for letters and postcards once you're in the country as they tend to go up frequently. Within Europe a stamp for a postcard presently costs 3 Kčs, for a letter 5 Kčs; sending mail overseas costs 6 Kčs for a postcard and 10 Kčs for a normal letter up to 10 grams. You'll find red letter boxes just about everywhere you look.

TELEPHONE

There are two different kinds of telephones in operation (provided they are not out of order) in the city. The first kind only accepts 1 crown coins and can therefore only be used for making local calls. Although the second type will accept 1, 2 and 5 crown coins, they are rather impractical for long-distance calls due to the fact you must constantly feed change into them. Taking this into account, if you want to make a long-distance call your best bet is to dial from either a post office or hotel: Bear in mind that hotels will charge a 20–30 percent commission for this service.

Important telephone numbers
Directory Enquiries: in Prague 120
International Enquiries: 0135 and 0149
Calling from abroad
International code + country code 42 + 2 (the code for Prague).

PARCEL DELIVERY SERVICE

DHL, the international parcel delivery service, offers a fast and reliable, but also expensive, means of sending parcels abroad.

The company has two offices in Prague:
City office (opposite the *Obecní dům*): Na poříčí, Prague 1. Tel: 267 525. Open: Monday–Friday 9am–6pm.
Main office: Běžecká 1, Prague 6, Strahov. Tel: 354 242 and 354 020, fax: 355 627 and 356 025. The latter is open until 6pm on Saturdays. A courier will also come and collect parcels.

TELEX, FAX & COMPUTER SERVICES

In most of the larger hotels you can both send and receive a fax or telex. In addition to this, many first-class hotels provide office services which include access to various computers and printers.

Telegrams can be sent from every post office, or over the phone (Tel: 127).

THE MEDIA

NEWSPAPERS, MAGAZINES, BOOKS

Foreign news publications are available at the kiosks located in hotels as well as in many bookshops. It has recently become possible to purchase German and English language weekly papers (published locally) in Prague and at various border-crossings.

The Tourist Information Centre stocks a variety of event calendars, restaurant guides and general information brochures in a number of different languages.

The English-language newspaper *The Prague Post* costs 5 Kčs and includes interesting articles on politics, culture and the economy as well as containing a programme of events and tips for restaurants and pubs.

International newspapers are sold by touts on Wenceslas Square and on the Old Town Square.

If you're looking for books written in German, French, Russian or English, your best bet is to make a trip to one of the international bookstores in Prague.

RADIO & TELEVISION

Radio and television are still under national control. The national radio channel 4 (medium waveband 255) has hourly news in both English and German. Radio 1 (the channel still has a tendency to float), organised and moderated by a young team in Prague, is one of the very first private radio stations to have emerged in the country.

There are three television channels: Channel 1 with programmes in Czech, evening news at 7.30pm; channel 2 CTV and Channel 3 which broadcasts in German, English and French and puts out satellite programmes such as CNN, SAT, RTL and MTV and the Sports Channel.

EMERGENCIES

MEDICAL AID

Many drugs are in short supply. Visitors should bring along any medication that they regularly require. Before setting out it is advisable to take out an international medical insurance. If you have to pay for treatment, make sure you get a receipt for money to be reimbursed, together with a certificate of the exchange rate at the time you pay.

Western visitors can be treated in the University Hospital. The first examination is free of charge; further treatment, operations and stays in hospitals must then be paid for in hard currency. For dental treatment there is also a special service for foreigners. Medication prescribed by Czech doctors can be paid for in crowns.

Fakultní poliklinika (University Hospital), Prague 2, Karlovo nám. 32. Tel: 299 381. Open: Monday–Friday 8am–4.15pm. Dentist: Monday–Friday 8am–3pm.
Dental Emergencies: Prague 1, Vladislavova 22. Tel: 261 374. Open: daily 7pm–7am.

Medical centre for diplomats and foreigners, Prague 5, Na homolce 724. Tel: 5292 2146. (Metro Station *Anděl* on the yellow line B, then further with bus No. 167 to the *Sídlistf Homolka* stop.) 1,000 Kčs must be deposited before treatment. Bring your passport.
Dental emergencies, Prague 1, Vladislavova 22. Tel: 261 374. Open: daily 7am–7pm.
Opticians: Oční optika, Na příkopě 13 and Vácslavské nam. 17 and 51.
Contact lenses: Oční optika, Mostecká 3 (Tel: 531 118) and Národní třída 37 (Tel: 221 071).

CHEMISTS

Chemists are open during normal business hours. In case of an emergency after hours, you'll find the address of the nearest chemist's open on emergency duty posted in the window.

The following chemists (lekárna) are open 24 hours a day:
Prague 1 (centre), Na příkopě 7. Tel: 220 081/2.
Prague 2 (Nové Město), Ječná 1. Tel: 267 181.
Prague 3 (Žižkov), Koněvova 150. Tel: 894 203.
Prague 4 (Nusle), Nám bratří Synků 6. Tel: 433 310.
Prague 5 (Smíchov), S. M. Kirova 6. Tel: 537 039.
Prague 6 (Břevnov), Pod Marjánkou 12.
Tel: 350 967.
Prague 7 (Letná), Obránců míru 48. Tel: 375 492.
Prague 8 (Libeň), Nám. dr. V. Holého 15.
Tel: 824 486.
Prague 9 (Vysočany), Sokolovská 304. Tel: 830 102.
Prague 10 (Vršovice), Moskevská 41. Tel: 724 476.

EMERGENCY NUMBERS

Lost and found: Prague 3, Olšanská; Prague 1, Bolzanova 5: Tel: 236 8887.
Emergency: Tel: 155
Ambulance: Tel: 373 333
Dental Emergency Service: Tel: 374
Fire Brigade: Tel: 150
Police: Tel: 158

CRIME

The crime rate in Prague is significantly lower than that of many Western European cities. However, since 1990 the incidence of offences like robbery, fraud and larceny have increased dramatically; cases of visitors having their handbags or wallets snatched are becoming increasingly common. It's therefore a good idea to deposit any valuables in the hotel safe and if possible to always park your car at a supervised car park.

In case of an emergency either consult with your hotel reception or contact the police directly by dialling 158. If you should have the misfortune to either lose or have your personal documents stolen, get in touch immediately with your embassy representative (*see Useful Addresses*).

GETTING AROUND

PIS, the Prague Information Service, provides tourists with all necessary information. However, one does have to be patient. It may take a while for you to get someone who speaks English, and even then it is not guaranteed that this someone will be able to answer your questions; he may have to go and ask someone else. This kind of scenario is repeated almost everywhere in Prague, whether you're trying to hire a car or buy a ticket.

The PIS main office is located in the Hradčanská Metro station; more centrally situated is the office on "The Moat": PIS, Na Příkopě 20. Tel: 544 444. In July 1992 the office on the Old Town Square still had no telephone; a further office is to be found on the B-level of the Main Station.

Detailed information can also be obtained from the travel agencies. Despite increasing competition from the private sector, **Čedok** remains the largest and most efficient agency. It maintains several offices in the city: Na Příkopě 18, tel: 212 7111 (information, international tickets, exchange bureau); Vácslavské nám. 55, tel: 227 096 (tours and stays in the spa towns).

Drivers can obtain service and information from: **Autoturist**, Opletalova 29. Tel: 223 544.

PUBLIC TRANSPORT

The various means of public transport are cheap and well synchronized. The network includes trams and buses, the Metro and the funicular up the Petřín Hill. Tickets can be purchased in shops, at the kiosks of the Prague Public Transport Executive, in restaurants, hotel receptions as well as the automatic ticket machines at the stops or stations. Tickets cost a flat rate of 4 Kčs. In the underground this allows you to travel for 60 minutes and to change as much as you like within that time, as long as you don't get out at a station; with buses and trams you have to cancel a new ticket every time you get in.

There is a special tourist ticket which allows an unlimited amount of journeys within a set period from the time off issue: for 2 days (40 Kčs); for 3 days (50 Kčs); for 4 days (60 Kčs); or for 5 days (70 Kčs). Transport on the entire network is completely free for children under 10 and adults over 70. Children under 16 only pay 1 crown. The 4 crown tickets must be cancelled inside the trams and buses and before entering the underground, where the cancelling machines are located directly at the entrance. There are no ticket or cancelling machines on the platforms, so even if you have no change make sure that you buy a ticket from the guard at the entrance and have it cancelled before you go through. For ticket sales and further information contact the information office of the Public Transport Executive in the Karlovo nám. metro station (yellow line B, exit to Palackého nám.). Tel: 294 682. Open: Monday–Friday 7.30am–3pm.

UNDERGROUND

The modern underground system links the centre with the suburbs and provides for convenient changes inside the city. It is a remarkably clean and quick means of public transport. The three lines have been developed with an eye towards expediency and by transferring it's possible to conveniently reach just about all the important tourist attractions located within the city.

The lines intersect at three main stations. From *Můstek* station at the bottom of Wenceslas Square you can take the green Line A over to the Lesser Quarter and Hradčany. The yellow Line B runs south to Charles Square and to the *Smichovské nádraži* station. The *Florenc* bus station and the northeast can be reached by travelling in the opposite direction. Line A intersects with the red Line C at the *Muzeum* station at the upper end of Wenceslas Square. The latter runs north to the main station, then to the *Florenc* bus station where it intersects with Line B before continuing to the terminus *Nádraží Holesovice*, the railway station for many of the trains on the Berlin–Budapest route. To the south it leads to Vyšehrad. Because of the high frequency of the trains (5–12 minutes), you hardly need to plan any more than about 30 minutes even for journeys into the suburbs. The Metro operates from 5am to midnight.

The red "M" signs outside the stations are small and look decidedly inconspicuous. But inside, the stations, often beautifully designed, are clean and clearly laid out. Network plans are prominently located at all entrances and above the platforms; the station you are at is highlighted; the stations you can change at are marked with the colour of the intersecting line.

Among the many tram and bus routes within Prague, Line 22 is probably the most interesting for visitors. It runs from Námestí Míru over Charles Square and along the Národní třída (National Street). It crosses the Vltava and then runs along the Karmelitská in the Lesser Quarter to the Lesser Quarter Square. From there it winds its way up the Castle hill and on along the Keplerova to the starting point for Strahov and the Petřín Hill. On line 22 it is possible to have an almost complete tour of the city for only 4 Kčs.

Prague has a comprehensive bus route network: Buses (autobus) and trolley buses (Trolejbus) run all day, particularly frequently in the suburbs, to connect with the Metro.

For information about national and international bus connections, tel: 221 445.

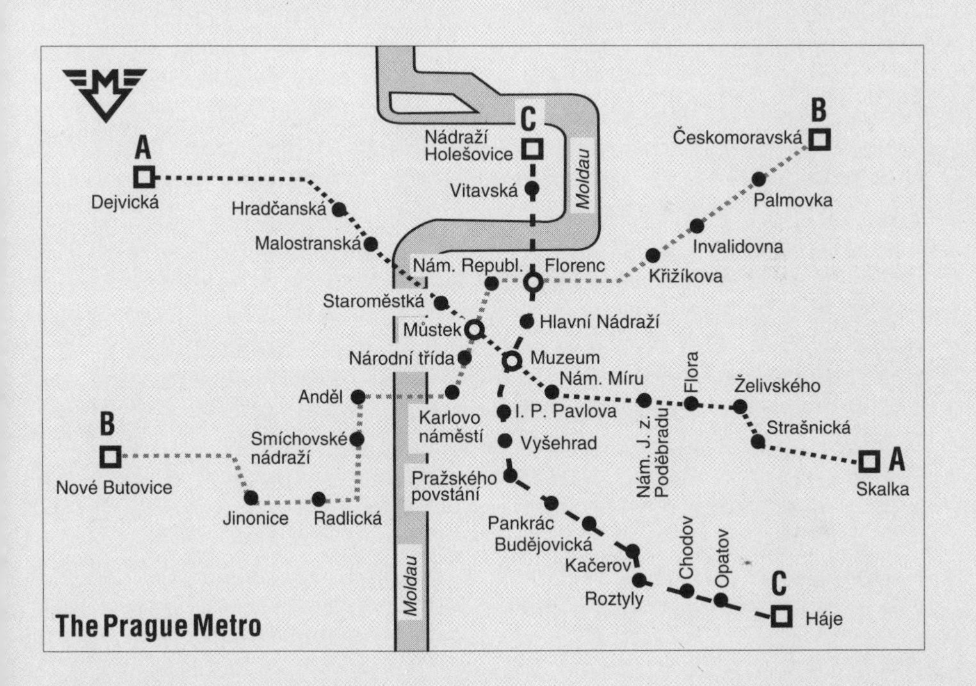

TAXIS

After midnight taxis constitute practically the sole form of public transport. You'll find a number of taxi stops in the city centre as well as in front of larger hotels. Even if the taxi driver seems reticent to do so, make sure the taximeter is running before setting off. Some taxis don't have a meter and in others it may be broken or set wrongly; in this case the driver will ask where you want to go and then quote a fixed price (in Summer 1992 this was 150 Kčs, three times the usual average rate).

During the day when you're not desperately marooned on some street corner, you can attempt to bargain or at least find a taxi whose meter is working. At nighttime this is difficult as prices are dictated by supply and demand and the driver knows the passenger is in the weakest position. Often there is no choice but to pay the sum demanded. Taxis can be ordered by dailing 201 941-5 and 202 951-5.

Drivers often turn out to be very friendly individuals with an intimate knowledge of the city. Many have other professions, but can earn more by driving taxis. They may be engaged for a half or entire day to drive passengers into and around the surrounding countryside. Having first established that the car is in relatively good order, the driver seems to know his way around and can perhaps even speak a bit of English, it pays to do a little bargaining until a mutually acceptable price is reached, anywhere between about £20 and £40. For couples and small groups this is certainly the most comfortable way of exploring the environs of the city. To get to the airport it's cheaper to take the ČSA Airport Taxi, as normal drivers would have to charge for the return journey as well.

TAXI STANDS IN THE CENTRE

Václavské nám. 38, Prague 1 (outside the Hvězda cinema). Tel: 224 095.
Václavské nám., Prague 1 (outside the fashion shop).
Václavské nám., Prague 1 (outside the shoe shop).
Národní třída, Prague 1 (Maj department store).
Municipal House (Obecní dům), Nám. Republiky, Prague 1.
Na příkopě , Prague 1 (outside the Slavic House).
Masaryk Station, Hybernská, Prague 1.
Main Station. Tel: 236 0402.
Old Town Square. Prague 1 (outside the Bohemian glass shop).
Královodvorská ul., Prague 1 (outside the *Vltava* travel agency).
Malostranské nám., Prague 1. Tel: 538 184.
Hradčany, Pohořelec, Prague 1. Tel: 538 531.
Klárov, Prague 1 (outside the *Malostranská* underground station).
Nábr. L. Svobody, Prague 1 (outside the University Hospital).
Karlovo nám. (Charles Square), Prague 1 (extension of the Žitná).

RENTAL CARS

Those wishing to go with Avis or Hertz can book from home as this generally works out a lot cheaper. To rent a car you have to be at least 21 years of age and in possession of a valid driver's licence. Credit cards are accepted.

Pragocar with its fleet of Škodas offers the cheapest terms. The Škodas not only have the advantage that Czech mechanics can repair them in their sleep, but also that they are not the most desirable of booty for car thieves. The Škoda costs about £15 a day and 15 pence per kilometre, or about £30 a day for unlimited mileage; for a weekend from 1pm on Friday to 9am on Monday they come at the bargain price of around £50 including 600 free kilometres. Because Pragocar is also represented in Carlsbad (tel: 017/22833-4), you can pick up a car in one town and deposit it in the other. This is not possible with other agencies.

Pragocar: Ruzyně Airport, tel: 36 8 707; Hotel Atrium, tel: 284 2043; Hotel Intercontinental, tel: 231 9595; Hotel Forum, tel: 419 0213; Opletalova Street 33, tel: 222 324; Pankráč, tel: 692 2875.
Czech Auto Rent: also offer cheap deals. Hotel Palace, Panská 12, tel: 236 1637; Hotel Prezident Nám. Curieových 100, tel: 231 4812 ext: 119.
Hertz: Ruzyně Airport, tel: 312 0717; Hotel Atrium, tel: 284 1111; Karlovo nám, tel: 290 122.
Avis: Hotel Atrium, tel: 284 2043.

TRAFFIC

Even for drivers who know Prague, the city can be a traffic nightmare. Large sections of the Old Town and the Lesser Quarter have been completely closed to traffic. Even if you manage to get through the maze of one way streets and cul de sacs to find yourself in the centre, at Wenceslas Square or the Powder Tower, you'll probably be turned away by the police (or may even get a ticket) unless you can prove you are resident at one of the noble hotels nearby. It is therefore highly advisable to leave your car at one of the supervised car parks and explore the city either on foot or by public transport.

PARKING

It is policy to keep the city centre as free of traffic as possible. To this end the centre has been divided into three zones: Zone A (Old Town), Zone B (New Town to the east of Wenceslas Square) and Zone C (to the west of Wenceslas Square). Within these zones parking spaces cost 1 crown for 30 minutes; outside they cost 1 crown for 60 minutes. Only cars of guests staying in hotels have access to Wenceslas Square and the surrounding streets. Guests obtain a permit at reception, which must be visibly displayed on the dashboard. This permit enables the guest to park in special slots reserved for the hotels or in the hotel garages. Vehicles that have been towed away are kept in Prague 10 (*Hostivař, Černokostelecká*), 15 km (9 miles) from the centre.

Locations of car parks in the centre: Platnéřská, Rytířská, Haštalská, Pařížská, Štěpánská, Národní třída, Náměstí Jana Palacha, Na Františku (outside the Československé airolinie), Pařížská (outside the Hotel Intercontinental), Gorkého náměstí, Opletalova (outside the main station), Politických vězňů, Malá Štěpánská, Těšnov (outside the Hotel Opera), Petrské náměstí (outside the Petrská věž Tower), Sázavská, Ibsenova, Škrétova, Tylovo náměstí (outside the Hotel Beránek).

BREAKDOWN

In case of breakdown ring the police number 158 or the tow-away service "Yellow Angel" on 154.

There are a number of 24-hour garages which will do quick repairs:
Prague 10, Limuzská 12, tel: 773 455; Prague 4, Macurova 1640, tel: 791 9157; Prague 8, Lodzská 14, tel: 855 8381.

Members of the AA or other automobile association can contact: Autoturist, Prague 1, Ječná 40. Tel: 293 723.

CAR REPAIRS

VW, Ford, Nissan: Severní XI, Prague 4. Tel: 766 752–4. Open: Monday–Friday 7am–3.45pm.
Fiat: Na strži 35, Prague 4. Tel: 692 2434. Open: Monday–Friday 6.30am–3.45pm.
Simca, Peugeot, Vauxhall Opel, Oltcit: Dáblická 2, Prague 8. Tel: 887 803, 888 257. Open: Monday–Friday 7am–3.45pm.
Vaz: Podbabská 3, Prague 6. Tel: 32 43 16, 32 40 78. Open: Monday–Friday 7am–3.45pm.
Daewoo, Renault, Dacia: Novostrašnická 46, Prague 10. Tel: 78215 01. Open: Monday–Friday 7am–3.45pm.
Škoda: Černokostelecká 114, Prague 10. Tel: 704 650. Open: Monday–Friday 6.30am–3pm.
BMW: Průbfěžná 76, Prague 10. Tel: 781 11 09. Open: Monday–Friday 8am–4.30pm.
Mercedes: Jeremiášova 11, Prague 5. Tel: 526 311, 523 229. Open: Monday–Friday 7.30am–5pm.

TYRE & BATTERY SERVICE

Vinohradská, Prague 2. Tel: 272 419. Open: Monday–Friday 6am–8pm.

PETROL STATIONS

24-hour service: Prague 3: Olšanská (special, super), Kališnická (special, diesel); Prague 4, Újezd u Pruhonic. Prague 5: Motol, Plzenská (special, super, diesel, unleaded). Prague 7: Argentinská (special, super, diesel, unleaded). Prague 8: Prosek (special, super, diesel); Českobrodská (special, super, diesel, unleaded).

Stations with **unleaded petrol**: Prague 3: Olšanská, Kališnická. Prague 4: Újezd u Průhonice, Podolská, U Pragocaru. Prague 6: Mackova, Ve struhách, Ruzyně (Dedina), Evropská, Bělohorská. Prague 7: Argentínská. Prague 8: Liberecká. Prague 9: Liberecká. Prague 10: Limuzská.

CAR WASH

Prague 1: SAO servis, Petrská 31. Tel: 231 5034. Prague 3: ČSAD Praha-Žižkov, Malešická 45. Prague 4: Bychl-Eurotechnik, Na strži 40. Tel: 692 2918. Prague 8: Pragoservis, Na Kundratce 19. Tel: 830 808. Prague 10: Autoservis, Černokostelecká. Tel: 701 931 (Line 58); Autoservis, Limuzská 12. Tel: 773 455.

FUNICULAR, VLTAVA CRUISES

The funicular up Petrín Hill (Újezd, Prague 5) runs daily from 5am–midnight. It can be reached by taking trams 9, 12 and 22. Prices and tickets of the public transport system are valid here as well.

The frequency of cruises on the Vltava depends on the weather. An evening cruise under the Charles Bridge is a particularly unforgettable experience. The trip lasts from 8pm–10.30pm and costs 550 Kčs inclusive of the evening meal supplied on board.

Further information can be obtained from the quayside on the Rašin bank at the Palacký Bridge.

WHERE TO STAY

Booking a hotel room in Prague particularly during the peak tourist season can prove to be a hopeless task. Due to the fact that hotel managements often have fixed contracts with foreign tour agencies, rooms can be reserved for travel groups for up to 14 days prior to the beginning of the intended trip. Because of this, independent travellers are frequently able to book a room only at the last minute, and even then, especially in the city, the number of beds available is quickly exhausted. A possible alternative to this often frustrating and sometimes fruitless searching is to reserve a room through a travel agency or the Čedok agencies, both of which have recourse to a certain number of rooms in various hotels. It is also possible, however, that you will not be able to find cheap accommodation through this channel.

HOTELS

The choice of hotels in Prague ranges from expensive luxury to cheap and simple. The expensive hotels are a category unto themselves; decor and service conform to international standards. As a rule the bill must be paid in Western currency. Credit cards are accepted.

Our list covers only some of the hotels typical for each category. Because of the high demand, it can be assumed that the choice will increase dramatically in the near future. We recommend the following:

A DELUXE (☆☆☆☆☆)

Alcron, Prague 1, Štěpánská 40. Tel: 235 9296.
Esplanade, Prague 1, Washingtonova 19. Tel: 222 552.
Inter-Continental, Prague 1, Náměstí Curieových. Tel: 2899.
Jalta, Prague 1, Václavské nám. 45. Tel: 265 541-9.
Palace, Prague 1, Panská 12. Tel: 236 0008.
Prague, Prague 6, Sušická 20. Tel: 333 8111.

A (☆☆☆☆)

Ambassador, Václavské náměstí 5. Tel: 221351-6.
Atlantik, Prague 1, Na poříčí 9. Tel: 231 8512.
Atrium, Prague 8, Pobřežní ul. Tel: 284 1111.
Diplomat, Prague 6, Evropská 15. Tel: 331 4111.
Forum, Prague 4, Kongresová ul. Tel: 410 111; Fax: 442/420684; Telex: 122 100.
International, Prague 6, Námfstí Družby 1. Tel: 321 051.
Olympic, Prague 8, Invalidovna, U Sluncove. Tel: 828 541.
Panorama, Prague 4, Milevská 7. Tel: 416 111.
Parkhotel, Prague 7, Veletržní. Tel: 20 380 7111.

☆☆☆

Axa, Prague 1, Na poříčí 40. Tel: 232 7234.
Belvedere, Prague 7, Obránců míru. Tel: 374 741.
Beránek, Prague 2, Bělehradská 110. Tel: 258 251.
Budovatel, Prague 1, Nám. Curieových 100. Tel: 231 4812.
Centrum, Prague 1, Na poříčí 31. Tel: 231 0135.
Družba, Prague 1, Václavská nám. 16. Tel: 235 1232.
Evropa, Prague 1, Václavské náměstí 25. Tel: 236 5274.
Flora, Prague 3, Vinohradská 121. Tel: 274 250.
Golf, Prague 5, Plzeňská 215a. Tel: 521 098.
Karl-Inn, Prague 8, Šaldova 54. Tel: 232 2551; Fax: 232 8030.
Koruna, Prague 1, Opatovická 16. Tel: 204 368.
Olympik II-Garni, Prague 8, Invalidnova, U Sluncové. Tel: 830 274.
Paříž, Prague 1, U Obecního domu 1. Tel: 231 2051.
Splendid, Prague 7, Ovenecká 33. Tel: 373 351-9.
Tatran, Prague 1, Václavské nám. 22. Tel: 235 2885.
U tří pstrosů, Prague 1, Dražického nám. Tel: 536 151.
Zlatá husa, Prague 1, Václavské nám. Tel: 214 3111

☆☆

Adria, Prague 1, Václavské nám. 26. Tel: 263 415.
Ametyst, Prague 2, Makarenkova 11. Tel: 259 256-9.
Balkan, Prague 5, Svornosti 28. Tel: 540 777.
Bohemia, Prague 1, Králodvorská 4. Tel: 231 3795-6.
Central, Prague 1, Rybná 8. Tel: 232 4351.
Erko, Prague 9, Kbely 723. Tel: 850 1138.
Hvězda, Prague 6, Na rovni 34. Tel: 368 965.
Hybernia, Prague 1, Hybernská 24. Tel: 220 431-2.
Juniorhotel, Prague 2, Žitná 12. Tel: 292 984.
Juventus, Prague 2, Blanická 10. Tel: 255 151.
Kriváň, Prague 2, I. P. Pavlova 5. Tel: 293 341-4.
Merkur, Prague 1, Těšnov 9. Tel: 231 6840.
Meteor, Prague 1, Hybernská 6. Tel: 235 8517.
Michle, Prague 4, Nuselská 124. Tel: 426 024.
Modrá hvězda, Prague 9, Jandova 3. Tel: 830 291.
Moráň, Prague 2, Na Moráni 15. Tel: 294 251-3.
Opera, Prague 1, Těšnov 13. Tel: 231 5609.
Ostaš, Prague 3, Orebitská 8. Tel: 272 860.

Praga, Prague 5, Plzeňská 29. Tel: 548 741-3.
Savoy, Prague 1, Keplerova 6. Tel: 537 450.
Transit, Prague 6, Ruzyňská 197. Tel: 367 108.
U blaženky, Prague 5, U blaženky 1. Tel: 538 286.
Union, Prague 2, Jaromírova 1. Tel: 437 858

BOTELS

Staying at one of the floating hotels along the Vltava in Prague can provide you with an experience you won't soon forget. However, they tend to be booked out by the coach tour operators:

Admiral, Hořejší nábřeží, Prague 5. Tel: 547 4.
Albatros, Nábřeží L. Svobody, Prague 1. Tel: 231 3634.
Racek, U Dvořecké louky, Prague 4. Tel: 425 793.

MOTELS

Club Motel Průhonice ☆☆☆☆, Tel: 723 241-9. In Průhonice southeast of Prague (15 minutes from the centre) on the motorway E 14 towards Brno; adjacent to the Botanical Gardens and the castle grounds of Průhonice. This hotel opened in 1991, with two sports halls, tennis and squash courts, bowling alleys, swimming pools, fitness centre etc.
Hotel Golf, Plzeňská 215a, Prague 5. Tel: 523 252-9. On the E 15 in Motol.

CAMPING

Information regarding camping is available through automobile associations or the headquarters of the Czech Automobile Association: **Autoturist**, Prague 1, Opletalova 29. Tel: 223 544-9. Open: Monday–Friday 9am–noon 1pm–4pm.

As a rule, campsites are divided into three categories; all have cold running water and lavatory facilities. Those falling into the third and simplest category "C" do not have access to electricity. In addition to plots reserved for caravans and tents, some campgrounds of the first category "A" also have holiday houses for rent. Generally speaking, don't expect anything luxurious with this kind of accommodation. It does however present a clean and inexpensive alternative to staying in a hotel:

Caravan, Prague 9, Kbely, Mladoboleslavská 27. Tel: 892 532. May–October.
Caravancamp, Prague 5, Plzeňská. Tel: 524 714. March–October.
Dolní Chabry, Prague 8 – Dolní Chabry, Ústecká ul. June–September.
Kotva, Prague 4, U ledáren 55. Tel: 461 712. May–September.
Mejto, Prague 10 – Nedvězí, Rokytná 84. Tel: 750 312-5. Open all year round.
Sportcamp, Prague 5, V podhájí. Tel: 521 802. March–October.
TJ Aritma, Prague 6, Nad lávkou 3. Tel: 368 351. April–October.
Xavercamp (bungalows), Prague 9, Božanovská 2098. Tel: 867 348. May–October.

PRIVATE LODGINGS

Private lodgings offer a comparatively inexpensive alternative to hotel rooms. There are a number of agencies through which you can book an apartment in the centre of Prague with prices starting at about £8 per day and person. It's also usually possible to find something suitable even at the last minute. Private people advertise accommodation on the street corners or at the exits to the motorways.

Private lodgings can be booked through the following agencies:

AVE (main station, upper hall), Prague 2, Wilsonova 8. Tel: 236 2560, 236 3075.
Čedok, Panská 5, Prague 1. Tel: 247 004, 225 657. Open: April–September weekdays 9am–9.45pm, Saturday 9am–6pm and Sunday 9am–4.30pm. From October to March the office is open on weekdays until 8pm and until 2pm at weekends.
Pragotour, U Prašné brány (near the Powder Tower), Prague 1. Tel: 231 7281 or 231 9245.

FOOD DIGEST

Apart from the pubs and wine bars (listed in separate sections), the city has many excellent restaurants. The following is a list of establishments in the different districts of the city where we found particularly good food and value for money.

THE LESSER QUARTER (Malá Strana)

Valdštejnská hospoda, (Waldstein Inn), Prague 1, Tomášská 16. Tel: 536 195. At the foot of Hradčany. Traditional decor, game specialities, not expensive. Open daily 11am–3pm, 6–10.30pm.
U tří pštrosů, Dražického nám. 12. Tel: 536 007. Next to the Charles Bridge. Traditional decor and Bohemian cuisine, upper price range. Open daily 11am–3pm, 6–11pm.
U čerta, Narudova 4. Tel: 530 975. Stylish, with waiters who speak English and German, reasonably priced. Open: daily noon–3.30pm, 6–11pm.
U Malfřů, Malézké nám. 11. Tel: 531 883. French restaurant since 1543, French prices, superb cuisine. Open: daily 9am–2am.
Nebozízek, Petřínské sady 411. Tel: 537 905. At the middle station of the funicular up Petřín Hill. Large

terrace, elegant restaurant with a view of the Vltava and the Old Town.

THE OLD TOWN (Staré Město)

U tři Gracii, Novotného lávka 5. Tel: 265 457. Moravian wines and Moravian cuisine.

Ve Skořepce, Skořepka 1. Tel: 228 081. Traditional atmosphere, reasonably priced. Open: Monday–Friday 11am–10pm, Saturday 11am–8pm. Closed: Sunday.

U sedmi andělu, Jilská 20. Tel: 266 355. Antique interior, middle of the price range. Open: daily except Monday noon–3pm, 6–11pm.

U Rudolfa II., Maiselova 5. Tel: 232 2671. The smallest of Prague's wine bars with good food based on traditional recipes. Affordable. Open: daily 10am–10pm.

Paříž, U Obecního domu (directly behind the Powder Tower). Tel: 232 2051. A beautifully restored restaurant with Art Nouveau furnishings. Upper price category.

Opera Grill, Karoliny Světlé 35. Tel: 265 508. International cuisine. Open: daily except Saturday and Sunday 7pm–2am.

U Sixtů, Celetná 2. Tel: 236 7980. Traditional Bohemian cuisine in a cellar. Open: daily noon–1am.

U zlaté Studny, Karlova 2. Tel: 220 593. Moravian wines and cuisine. Open: daily 11am–3pm, 5pm–midnight.

THE NEW TOWN (Nové Město)

Adria, Národní třída 40 (on Jungmann Square). Tel: 62637. Summer terrace but still reasonable. A café adjacent. Open: daily 10am–10pm.

Volha, Myslíkova 14. Tel: 296 406. Excellent cuisine from the Black Sea/Caucasus region. Middle of the price range. Open: daily except Saturday and Sunday 11am–midnight.

Vltava, Rašinovo nábřeží (right bank of the Vltava where the cruisers pull in). Tel: 94964. Terrace on the Vltava and a cosy room for bad weather. Reasonably priced and generous portions. Open: daily 11am–10pm.

INTERNATIONAL SPECIALITIES

Alex, Prague 1, Revoluční 1. Tel: 231 4489. German specialities. Open: daily 11am–1am.

ASIA, Letohradská 50, Prague 7. Tel: 370 215. Asian cuisine. Open: 11am–11pm except Saturday and Sunday.

Berjozka, Prague 1, Železná 24. Tel: 223 822. Russian specialities. Open: Monday–Saturday 11am–11pm.

Čínská restaurace, Vodičkova 19, Prague 1. Tel: 262 697. Chinese cuisine. Open: daily noon–3pm, 6am–11pm, except Sunday.

Gruzia, Prague 1, Na příkopě 29. Specialities from Gruzinskaya. Open: daily 11am–7pm and 6pm–1am.

Habana, Prague 1, V Jámě 8. Tel: 260 164. Specialities from the Caribbean and Cuban cocktails. Open: daily noon–4pm and 6pm–midnight, except Sunday.

Jadran, Prague 1, Mostecká 21. Tel: 534 671. Balkan specialities and wines. Open: daily 11am–10pm.

Jewish Restaurant, Prague 1, Maiselova 18.

Mayur, Prague 1, Štěpanská 61. Tel: 236 9922. Indian specialities. Open: daily noon–4pm, 6–11pm, except Sunday.

Pampa, Prague 6, Karlovarská 1/4. Tel: 301 7731. Argentinian restaurant. Open: daily noon–4pm, 5.30pm–midnight, except Sunday.

Peking, Legerova 64, Prague 2. Tel: 293 531. Chinese cuisine. Open: daily 11.30am–3pm, 5.30–11pm, except Sunday.

Pelikán, Prague 1, Na příkopě 7. Tel: 220 782. Elegant surroundings in the pedestrian precinct. Open: daily 11am–11.30pm.

Praha Expo 58, Prague 7, Letenské sady 1500. Tel: 377 339. Beautiful view of the Vltava and the Old Town; terrace café in the summer. Open: daily 1.30–3pm, 6–11pm.

Rostov, Prague 1, Václavské nám. 21. Tel: 262 469. Bohemian specialities, tables outside in the summer. Open: daily 1–3.30pm, 5–11pm.

Rotisserie, Prague 1, Mikulandská 6. Tel: 206 826. The menu includes a variety of steaks. Open: daily 11.30am–3.30pm, 5.30–11.30pm, except Sunday.

Savarin, Prague 1, Na příkopě 10 (in the arcade). Tel: 22 20 66.

Sofia, Prague 1, Václavské nám. 33. Tel: 264 986. Balkan specialities. Open: daily 11.30am–3pm, 6–11pm.

Thang Long, Dukelských hrdinů 48, Prague 7. Tel: 806 541. Vietnamese cuisine. Open: daily noon–3pm, 5–11pm.

Trattoria Viola, Národní 7, Prague 1. Tel: 266 732. Italian cuisine. Open: daily 11.30am–3pm, 5.30–11pm. Closed: Saturday and Sunday during July and August.

BOHEMIAN SPECIALITIES

Barrandov Terraces, Prague 5, Barrandovská 171. Tel: 545 309, 545 409.

Černý kůň (Black Horse), Prague 1, Vodičkova 36. Tel: 262 697. Traditional Prague cuisine, Pilsner beer. Open: daily 11am–1pm.

Halali-Grill, Prague 1, Václavské nám. 5. Tel: 221 351. Game dishes.

Hanavsky Pavillon, Prague 7, Letenské sady 173. Tel: 325 792. Open: daily 8.30pm–12.05am, terrace in summer noon–8.30pm.

Lví dvůr (Lion's Den), Prague 1, U prašného mostu 6, (in the Prague Castle complex). Tel: 535 386. Open: daily 10am–5pm, except Monday.

Myslivna, Prague 3, Jagellonská 21. Tel: 277 416. Game specialities. Open: daily 11am–11pm.

Obecní dům (Municipal House), Prague 1, Nám. Republiky 1090. Tel: 231 9754. French restaurant

in Art Nouveau, Pilsner Bier. Open: daily 11am–11pm.
Rybárna (fish restaurant), Prague 1, Václavské nám. 43. Tel: 227 823. Open: daily 11am–10pm, except Sunday.
Savarin, Prague 1, Na příkopě 10 (in the arcade). Tel: 222 066.
Slavie, Prague 1, Narodní 1. Tel: 265 760. Nextdoor to Prague's most famous coffee house. Open: daily 11am–11pm.
Slovanský dům (Slavic House), Prague 1, Na příkopě 10. Tel: 224 851. Largest restaurant complex in Prague. Open: daily 11am–midnight, except Sunday.
Theatre Restaurant, Prague 1, Národní 6 (National Theatre). Open: daily 11am–4pm, 5pm–midnight.
U kalicha (The Chalice), Na Bojišti 12, Prague 2. Tel: 290 701. Open: daily 11am–11pm. A venerable institution. Many tourists.
U krále brabantského (The King of Brabant), Prague 1, Thunovská 15. Tel: 539 975. A good stock of wine. Open: daily 1–11pm, except Sunday.
U Lorety, Prague 1, Loretanské nám. 8. Tel: 536 025. Elegant Prague restaurant opposite the Loreto Church; garden terrace in the summer. Open: daily 11am–3pm, 6–11pm.
Vikarka, Prague 1, Vikářská 6. Tel: 536 497. Famous pub at the Castle, popular amongst Czech artists.
Vysočina, Prague 1, Národní třída 26. Tel: 225 773.

DRINKING NOTES

BEER

What else, you may ask, is there to do at inns and pubs besides drink beer? The inhabitants of Prague don't even bother to pose this question at all! If you believe statistics, each resident of the "Golden City on the Vltava" downs about 150 litres of beer per year. The usually hopelessly over-crowded pubs and inns attest to the fact that while a great quantity of the brew is certainly consumed in the city, not much of it is drunk in privacy at home.

Czech beer is probably the finest beer in the world. Its quality is largely thanks to the famous "Bohemian hops", which have been culivated in Northern Bohemia ever since the Middle Ages. The hop centre is Žatec (Saaz). In Prague both light (*svetlé*) as well as dark (*tmavé*) beer is poured. The degrees (°) do not refer to the alcohol content but the percentage content of the original wurt. Light draught beer (10°) has an alcohol content of between 3 and 4 percent, lager (12°) five percent. The stronger, dark varieties (13° and more) are comparable with the German strong beers.

The most famous beers are *Pilsener Urquell* from Plzeň (Pilsen) and *Budvar* from České Budějovice (Budweis). But the beer from Prague's Smíchov brewery is also very good, and then there is the strong dark beer brewed on the premises at U Fleků. Another strong beer comes from the Prague district of Braník.

BEER HALLS

U dvou kocek (The Two cats), Prague 1, Uhelný trh 10. Tel: 267 729. One of the most popular of Prague's pubs: Pilsner Urquell.
U Fleků, Prague 1, Křemencova 11. Tel: 293 246. The malthouse and brewery date from 1459. 13° dark beer still brewed on the premises, accompanied by traditional Prague cabaret.
U medvídků (The Bears), Prague 1, Na Perštyně 7. Tel: 235 8904. This traditional restaurant offers Southern Bohemian and Old Czech specialities, accompanied by 12° Budvar.
U Pinkasů, Prague 1, Jungmannovo nám. 15. Tel: 261 804. A popular pub that has been pulling Pilsener Urquell since 1843.
U supa (The Vulture), Prague 1, Celetná 22. This pub dates back to the 14th century and serves Braník Special.
U sv. Tomáše (St Thomas's), Prague 1, Letenská 12. Tel: 530 064. 12° beer from Braník.

FURTHER BEER INSTITUTIONS

Bránický sklípek, Prague 1, Vodičkova 26. Tel: 260 005. 14° beer from Braník.
Černý Pivovar (Black Brewery), Prague 2, Karlovo nám. 15. Tel: 294 451. 12° Pilsener Urquell.
Na Vlachovce, Prague 8, Rudé armády 217. Tel: 840 576. 12° Budvar.
Plzeňský dvůr, Prague 7, Obránců míru 59. Tel: 371 150. 12° Pilsener Urquell.
Rakovnická pivnice, Prague 5, S.M. Kirova 1. Tel: 542 531. 12° Bakalar beer from Rakovník.
Smíchovský sklípek, Prague 1, Národní 31. Tel: 268 172. 12° beer from Smíchov.
U Bonaparta, Prague 1, Nerudova 29. Tel: 539 780. 12° beer from Smíchov.
U Černého vola, (The Black Bull), Prague 1, Loretánské nám. 1. Tel: 538 637. 12° beer from Velké Popovice.
U dvou srdcí (Two Hearts), Prague 1, U lužického semináře 38. Tel: 536 597. 12° Pilsener Urquell.
U Glaubiců, Prague 1, Malostranské nám 5, 12° beer from Smíchov.
U Schnellů, Prague 1, Tomášská 2. Tel: 532 004. 12° Pilsener Urquell.
U Sojků, Prague 7, Obránců míru 40. Tel: 379 107. 12° Pilsener Urquell.
U zlatého tygra, (The Golden Tiger), Prague 1, Husova 17. Tel: 265 219. 12° Pilsener Urquell.

WINE BARS

Every bit as popular as the pubs are the wine bars, known as *vinárna*, which serve predominantly Czech wines. The best wine comes from Žernoseky in the Elbe Valley, where the Melník wines are also cultivated. Good wines are also produced in Southern Moravia, in places like Mikulov, Hodonin, Znojmo or Valtice. Some wine bars serve wine from "their own" cooperatives. At one time there used to be vineyards even in Prague; a fact recalled by the name of the district Vinohrady above Wenceslas Square.

Foreign wines from Yugoslavia or Bulgaria are served in their respective speciality restaurants. Chianti Ruffino is served in the Trattoria Viola.

Snacks are also served in the wine bars. Especially in the Lesser Quarter, in recent years many wine bars have fallen into private hands and become expensive restaurants; some have thus lost much of their original charm.

Opening times vary, but most establishments stay open until midnight, some to 3am. The wine bars are often the last places to go for night owls in Prague.

Blatnice, Prague 1, Michalská 8. Tel: 224 751. Moravian wines from the area around Blatnice. Open: 11am–11pm, except Saturday.

Klašterní vinárna (Monastery Wine Bar), Prague 1, Národní 8. Tel: 290 596. This large wine bar is built in the walls of the former Ursuline Convent, and serves wines from Moravia and Nitra. Open: daily 11am–1am.

Lobkovická vinárna, Prague 1, Vlašská 17. Tel: 530 185. A historic wine bar in the Lesser Quarter, dating from the 19th century; wines from Melník are served here.

Makarská, Prague 1, Malostranské nám. 2. Tel: 531 573. Balkan specialities and wines. Open: daily 11.30am–10.30pm.

Nebozízek, Prague 5, Petřínské sady. Tel: 537 905. Reached by funicular from the Lesser Quarter; with an impressive view of Prague and Hradčany. Open: daily 11am–6pm, 7pm–midnight.

Parnas, Prague 1, Smetanovo nábřeží 2. Tel: 265 017. International cuisine with an unforgettable view of the Hradčany. Open: daily 7pm–1am, except Sunday.

Svatá Klara (St Clare), Prague 7, U trojského zámku 9. Tel: 841 213. Exclusive cellar wine bar at the entrance to Prague Zoo. Open: 6pm–1am, except Sunday.

U labutí (The Swans), Prague 1, Hradčanské nám. 11. Tel: 539 476. Exclusive wine bar near the castle serving South Moravian wines. Open: 7pm–1am.

U malířů (The Painters), Prague 1, Maltezské nám 11. Tel: 531 883. A typical Old Prague wine bar dating back to 1583. Open: 11am–3pm, 6–11pm, except Sunday.

U mecenáše (The Sponsor), Prague 1, Malostranské nám. 10. Tel: 533 881. There was an inn in the house "at the sign of the Golden Lion" as long ago as 1604; today this wine bar is among the nicest in the city. Open: 5pm–1am, except Saturday.

U patrona (The Patron), Prague 1, Dražického nám. 4. Tel: 531 661. Cosy atmosphere, South Moravian wine. Open: Monday–Friday 4pm–midnight.

U pavouka (The Spider), Pragl, Celetná 17. Tel: 231 8714. This historic wine bar with its Gothic and Renaissance halls serves wines from Southern Moravia. Open: 11.30am–3pm, 6–11.30pm, except Sunday.

U plebána, Prague 1, Betlémské nám. 10. Tel: 265 223. First-class cuisine with excellent wine from Znjomo. Open: 7am–11pm, except Sunday.

U zelené žáby (The Green Frog), Prague 1, U radnice 8. Tel: 262 815. This venerable institution has poured wine from Velké Žernoseky in Bohemia since the 15th century.

U zlaté hrušky (The Golden Pear), Prague 1, Novy svět 3. Tel: 531 133. A stylish bar in the Romantic atmosphere of the Castle. Open: daily 6.30pm–12.30am.

U zlaté konvice (The Golden Pot), Prague 1, Melantrichova 20. Tel: 262 128. Wine bar in cellars whose walls date from the 14th century; wines from Valtice.

U zlatého jelena (The Golden Stag), Prague 1, Celetná 11. Tel: 268 595. Cellar wine bar with wines from Southern Moravia. Open: Monday–Friday noon–midnight, Saturday–Sunday 6pm–midnight.

FURTHER WINE BARS

Beograd, Prague 2, Vodičkova 5. Balkan wines, restaurant.

Fregata, Prague 2, Ladova 3.

Mělnícká vinárna, Prague 1, Národní tr. 17. Wines from Melník.

Slovácká vícha, Prague 1, Michalská 6. Bzenec wine.

U Golema, Prague 1, Maiselova 8.

U markýze (The Marquis), Nekázanka 8, Prague 1.

U Šuterů, Prague 1, Palackého 4.

COFFEE HOUSES

Read, smoke, discuss, let the day go by; that's what the coffee houses are all about. Coffee houses still offer more than simply coffee and cake, and have little in common with their cousins in Germany or Austria, which are more meeting places for ladies of advancing years. The Prague coffee house remains an important element in the daily life of the city. The coffee on offer, however, often prepared in ten or more variations, is not always the strongest.

Arco, Prague 1, Hybernská 16.

Columbia, Prague 1, Staroměstské nám. 15.

Evropa, Prague 1, Václavské nám. 29.

Kajetánka, Prague 1, Hradčany, Kajetánska zahrada.

Malostranská kavárna, Prague 1, Malostranské nám. 28.

Mysák, Prague 1, Vodičkova 31.

Obecní dům, Prague 1, Náměstí Republiky.

Praha, Prague 1, Václavské nám. 10.

Savarin, Prague 1, Na příkopě 10.

Slavia, Prague 1, Národní 1.

U zlatého hada, Prague 1, Karlova 18.

THINGS TO DO

It is not difficult to find your bearings in Prague, especially as the most important sights can be reached on foot. The city's small centre (Prague 1) is divided into the historic quarters of Malá Strana (Lesser Quarter), Staré Město (Old Town), and Nové Město (New Town). The latter is centred around Wenceslas Square and extends to the road Na příkopě. Adjacent and to the north is the Staré Město, which extends across the Old Town Square (Staroměstské náměstí) and the right bank of the Vltava and the Charles Bridge. The picturesque Malá Strana lies on the left of the river. Two other self-contained districts are the Josefov (Jewish Quarter) and Hradčany, the Castle Quarter.

It's a good idea to allow yourself at least three days for a visit to Prague. The following daily tour suggestions have been made to aid visitors who are only in Prague for short stay.

DAY 1

We suggest that your first day's sightseeing could be spent strolling through the winding streets and alleyways of the Old Town to the Jewish Cemetery and the Old Town Square. The day can be ended not far from the Powder Tower with a classy dinner at the Hotel Paříž.

The tour commences at Wenceslas Square and leads via the Na můstku right into the Old Town, and the main sights to be seen here are as follows: The Flea Market – Old Town Hall – Old Town Square with the Jan Hus Monument – Karlova ulice – Clam-Gallas Palais – Husova ulice – Bethlehem Chapel – Náprstkova ulice – Bank of the Vltava with the Smetana Museum – Old Town Bridge Tower and the Crusaders' Church – Clementinum and the Church of St Saviour – Platnéřska ulice – Maiselova ulice – Jewish quarter with the old Jewish Cemetery, Klaus Synagogue, Old-New Synagogue and the Jewish Town Hall.

The visitor can now pause for lunch and has the choice between the Kosher restaurant in the former lower council chamber of the Town Hall, the wine bar U Golema (Maiselova 8) or the restaurant U Barona (Pařížská 19). After lunch go back along the Pařížská to the Old Town Square with the Church of St Nicholas, Palais Kinsky and the Týn Church. Then into Štupartská ulice – Church of St James – Celetná ulice – Powder Tower – Municipal House. Reward yourself at the end of a long day of sightseeing with a sumptuous dinner in the Paříž Restaurant (tel: 232 2051) next door.

DAY 2

The second day of sightseeing might lead you over the Charles's Bridge to the Lesser Quarter and up to Hradčany. This day's tour could be brought to a close in a typical Lesser Quarter pub serving both beer and meals.

The starting point is at the Old Town Bridge Tower. Having walked across the Charles Bridge, you'll be greeted by the Lesser Quarter Bridge Towers. The tour continues into the Lázeňská ulice – the Church of St Mary in Chains – Velokopřevorské náměstí – Island of Kampa – the Maltese Square with the Nostiz Palais (housing the Ministry of Education and the Arts), and into Karmelitská ulice with the Vrtbovsky palác (No. 25). Along Mostecká ulice (Bridge Street) and to the Lesser Quarter Square with the Church of St Nicholas. Here it's time for a break, before the walk up to the castle. There are the two pubs, the U Glaubicû and the U Schnellû, or the café in the middle of the square.

Then into the Neruda Alley and up to the Hradčanské náměstí and on: Loretánská ulice – Loreto Shrine – Novy Svět – back to Schwarzenberg Palais – Archbishops' Palace – Prague Castle with St Vitus' Cathedral and the Royal Palace – St George's Basilica – Mihulka Tower – Golden Alley – Lobkovic Palais – Black Tower – Old Castle Steps – Valdštejnská ulice. And there we reach the Valdštejnská Hospoda (Tomášská 16, tel: 536 195) our tip for a hearty Bohemian meal. If you don't get a place there, we suggest you go down the Tomášská to the Lesser Quarter Square, turn left at the corner to arrive at the venerable pub U Svatého Tomáše in Latenská 12, (tel: 530 064).

DAY 3

The third day's excursion could take you across Wenceslas Square and the New Town, with the National Theatre, the Vltava Quay and Charles Square. Dinner is a choice of either good, solid Bohemian food in U Fleků, or refined Russian dining in the Volha.

The tour begins at the equestrian statue of St Wenceslas in front of the National Museum and continues down Wenceslas Square past the classic Art Nouveau hotels – Evropa, Zlatá Husa, the Ambassador, the Alfa Palais (No. 28) and the Peterka House (No. 12) to the Koruna Palace at the end. Do an about-face and amble back along the other side of the square until you've reached about the half-way point, and then left into the Jindrísska ulice with the Hotel Palace; then along the Panská ulice – Na příkopě – Powder Tower and back to Wenceslas Square. Then continue into the Jungmannovo náměstí with the Church of St Mary of the Snows – Národní třída with the Maj department store – Kanka House (No. 16) and the Church of St Ursula (No. 8). For lunch, the monastery wine cellar offers a selection of tasty wines and dishes, or failing that there is the pub U Medvídků just around the corner (Na Perštyně 7, opposite the Maj department store).

The end of the Národní třída is completely dominated by the National Theatre. The Café Slávia stands opposite. Now walk upstream along the bank of the Vltava. Our route: Slavic Island – Mánes House – Jiráskův Bridge where the Vltava cruisers pull in and where there is a small pub with terrace called "Vltava" – Resslova ulice – the Church of St Wenceslas – the Church of SS Cyril and Methodius Church – Charles Square with the "Faust House"; the Church of St John on the Rocks is located to the south, opposite the Emmaus Monastery; back via the Church of St Ignatius and the New Town Hall which provides a counterpoint to the Faust House.

Especially in the late afternoon the U Fleků is the perfect place to sit back and take the load off your feet (turn left by the Town Hall into the Myslíkova and take the second right into the Křemencova). But for those who don't like the noise here, perhaps a better choice is the Volha (Myslíkova 14, tel: 296 406. Open: 11am–12 midnight).

TOUR GUIDES & INTERPRETERS

Čedok agencies and PIS all offer foreign language-speaking tour guides to travel groups as well as to independent travellers for day excursions or trips lasting several days. They can also furnish you with an interpreter or translator upon request. It is not unusual in Prague to "rent" a tour guide who then accompanies you on foot through the city. The price for an hour of this particular kind of service is about 70 Kčs per person.

OTHER ATTRACTIONS

Apart from visiting its historical sights, there are plenty more interesting things to see in Prague. Here are some proposals:

Prague Zoo is situated in Prague 7, Troja, and can be reached by the Metro line C, getting out at the station Nádrazí Holesovice and continuing on the 112 bus. A visit to the attractive zoo can also take in the nearby Troja Castle. Founded in 1931, today the zoo is home to 2,000 animals of 600 species. The breeding of the famous Przewalski horse is of international importance: Prazská zoologická zahrada, Prague 7 Troja, U Trojského zámku 3. Opening hours: October–March 7am–4pm; April 7am–5pm; May 7am–6pm; June–September 7am–7pm.

The Botanical Gardens are part of the university and are to be found in the Nové Město not far from the Church of St Nepomuk on the Rocks: Botanická zahrada, Prague 1, Na slupi. Open: 7am–7pm.

Prague Observatory, Prague 1, Petřín 205 (next to the the cable car station on the Petřín Hill. Open: daily except Monday: January, February, October, November, December 6–8pm; March, September 7–9pm; April, August 8–10pm; May, June, July 9–11pm.

CITY TOURS

Many agencies including the state travel agency Čedok not only organise tours of the city but cultural events as well. The 3-hour city tour "Historic Prague" is run throughout the year. Departure points are the Čedok bus park in the Bílkova 6 (opposite the Intercontinental Hotel) and the hotels Panorama, Forum and Atrium.

PRIVATE TOURS

Private tours of the city are organised by the Prague Information service (PIS), as well as travel agencies AVE and Pragotour (*see Tourist Information*).

Hradčany: Tours of Hradčany are organised by: Informacní stredisko prazského hradu, Prague 1 (Hradcany), Vikárska 37 (on the northern side of St Vitus' cathedral). Tel: 2101/3368.

EXCURSIONS

Čedok organises a number of day-long excursions into the environs of Prague. The palette includes: Castles in Bohemia, Southern Bohemia, Gothic architectural jewels in Bohemia, Attractive Central Bohemia, Vineyards of Bohemia and cruises on the Vltava. The buses depart from the Panorama Hotel at 8am, the Forum Hotel at 8.10am, the Atrium Hotel at 8.20am and from Čedok, Bílkova 6 at 8.40am. Tickets may be purchased from the hotel receptions as well as from Čedok: Prague 1, Bílkova 6, tel: 231 8855; Na Příkopě 18, tel: 212 7111; Prague Ruzyně Airport, tel: 367 802.

For those with their own car there is a host of attractions in the vicinity of the capital, all of which are easy to reach. Here is a selection:

Průhonice: Renaissance palace with botanical garden, reached via the motorway to Brno. The palace contains one of the largest herbarium collections in the world.

Konopiště: Palace with rose garden, bathing pool and English-style park. The original 14th-century castle was converted in the 18th century and served as a hunting lodge for the Austrian archduke Franz Ferdinand until his assassination in Sarajevo. There is a weapons collection with over 5,000 pieces.

Slapy Dam: A popular summer excursion for the people of Prague, located in the Vltava valley to the south of the city.

Karlštejn: Built by Charles IV in 1348 as the representative residence of the king and the depository of the crown jewels. Situated to the southwest of Prague, it is considered to be one of the most beautiful castles in Europe. It can be reached from the motorway in the direction of Plzeň as far as Baroun.

Koněprusy: Caverns near Karlstejn castle, discovered in 1950. Remains of prehistoric man as well as a coin forging workshop dating from the 15th century.

Křivoklát: 15th century castle set in an extensive area of forest which has been adopted by UNESCO in the programme "Man and the Biosphere". In the summer there are performances of music and theatre.

Rakovník: One of the oldest towns in Central Bohemia in the middle of the hop growing area. Its historic centre contains a number of important monuments.

Kladno: Historic mining town to the northwest of Prague (past Ruzyně Airport). There is a baroque palace with a museum of mining.

Lidice: Situated between Prague and Ladno, a memorial to Nazi massacre of 1942 in which all menfolk were shot and women rounded up and taken away to their fate at Ravensbrück concentration camp. The surviving children were dispersed through Germany to be renamed and raised as Germans. In an action that formed part of the Germans' brutal reprisals for the assassination of the deputy leader of the SS, Reinhard "the hangman" Heydrich, by the Resistance, the village was burned to the ground.

Veltrusy: Baroque Palace with original interior. It contains a collection of Asian porcelain, crystal and tapestries. It can be reached by following the E 55 in the direction of Teplice.

Nearby is the village of **Nelahzeves** with the house in which Antonín Dvořák was born.

Mělník: A wine producing town at the confluence of the Vltava and the Elbe, whose origins go back to the 9th century.

Mladá Boleslav: Home of Škoda cars. Přemyslid castle dating from the 10th century. Ancient centre containing interesting examples of avant-garde blocks of flats from the 1930s. It lies to the northeast of Prague along the E 65.

Přerov nad Labem: Open air museum of vernacular architecture. Located to the east of Prague on the E 67 towards Poděbrady.

Kutná hora (Kuttenberg): In the 13th and 14th century this town became the economic centre of Bohemia on account of its silver mines, and was the site of the royal mint. The Brass Music Festival "Kmochs Kolín" takes place here in June. The imposing Church of St Barbara was designed by Peter Parler. It is situated some 10 miles to the southeast of Kolín.

Český Sternberk: One of the best preserved castles in the country, dating from the 13th century. Its location on a clifftop above the River Sazava made it impregnable for centuries. Musical performances are held here in the summer. It lies to the southeast of Prague along the E 50.

CALENDAR OF EVENTS

A visit to Prague could be combined with a visit to one of the countless festivals or sporting events that take place in the country throughout the year. Here is a selection (approximate dates only):

December/January, Prague: Prague Winter – a theatre and music festival.

April, Prague: Interkamera – an international photography and technology exhibition.

May/June, Prague: Prague Spring Music Festival.

June, Prague: Concertino Praga – debut performances from a variety of young, talented musicians.

June, Kolín/Central Bohemia: – the International Brass Music Festival "Kmochs Kolín".

June (end of the month), Strakonitz/Bohemian Forest: Folklore Festival.

June/July, Stražnice/Southern Moravia: International meeting of different folk groups. in ethnic costume, celebrated by traditional dancing and music.

July, Karlovy Vary: International Film Festival – the traditional forum for Eastern European films.

July, Chrudim/Eastern Bohemia: Puppet Festival.

July, Znojmo/Southern Moravia: Royal Festival – medieval tournaments and events.

July, Zelezný Brod/Eastern Bohemia: Folklore Festival.

August, Domažlice/Western Bohemia: Chode Festival, a festival of the culturally independent border people with dancing, music and even bagpipes.

August, Mariánské Lázně: Chopin Music Festival.

September (the middle of the month), Žatec/Northern Bohemia: Žatec Hops Festival.

September/October, Namest na Hane/Northern Moravia: Harvest Festival.

October, Pardubice/Eastern Bohemia: Big international steeplechase – one of the most gruelling and notorious steeplechase competitions in the world.

AVANT-GARDE ARCHITECTURE IN PRAGUE

1. Hotel Evropa, built 1900, Art-Nouveau style.
2. Peterka House, architect Jan Kotěra. Built 1900, Viennese influence, Art Nouveau.
3. Urbánek House, designed by Jan Kotěra. Built 1912, Bohemian characteristics.
4. The Stenc House, designed by Otakar Novotný.
5. Hubschmann's apartment house, built 1911.
6. Kotěra office block, built 1924.
7. 1913 street lamp and seat, by Vladislav Hofmann.
8. House of the Black Madonna, Josef Gočár, 1912.
9. Rondo-Cubism office block, Pavel Janák, 1922.
10. Office block, also by Pavel Janák, 1923.
11. Hostel, designed by Novotný.
12. Mánes House of Artists, also by Novotný, 1928.
13. Black Rose (shopping arcade), 1929.
14. Hotel Juliš, 1933.
15. Lindt House, 1927.
16. MAJ Department Store, designed in 1968.

CULTURE PLUS

Prague has a very rich and varied cultural palette. More difficult than the choice is the problem of obtaining tickets. Sometimes you can book through Čedok, but this usually has to be done days in advance. Even then, it isn't guaranteed that you'll get tickets.

Often the only alternative is to go directly to the box office, where you might read *vyprodáno* – sold out. But don't let that put you off. If you are polite and point out that you came all the way just to see it then there's a good chance that you'll be successful. Some performances are reserved for companies and cooperatives. But it is still worth going along and looking around on the off chance.

TICKETS

Tickets for concerts and other public events can be obtained from the following booking offices:

Sluna: Panská 4, Passage Černá růže. Tel: 265 121, 221 206. Open: Monday–Friday 10am–6pm. Concerts and theatre: Václavské náměstí 28, Alfa Arcade. Tel: 260 693. Open: Monday–Friday 10am–6pm.

PIS – Prazská informacní sluzba (Prague Information Service): Staroměstské nám. (Old Town Square) 28. Tel: 224 453. Open: Monday–Friday 8am–6pm, Saturday 9am–6pm. Mainly concerts.

BTI – Bohemia Ticket International: Na příkopě 16. Tel: 227 838; Václavske nám. 25. Tel: 260 333; Karlova 8; Salvátorská 6. Tel: 231 2030. Open: Monday–Friday 9am–noon, 1–6pm; Saturday and Sunday 9am–5pm.

Czech Philharmonic: Prague 1, Masná 21. Tel: 231 9164. Open: Monday–Friday 1.30–6pm.

Symphony Orchestra FOK: Prague 1, U obecního domu 2. Tel: 232 5858. Open: Monday–Friday 9am–noon, 1.30–4.30pm (Friday 3pm).

The box office for the **National Theatre** and the **New Theatre** is situated in the glass building of the New Theatre, Národní třída 4 (Monday–Friday 10am–6pm, Saturday and Sunday 10am–noon). Prices for theatre and opera tickets are between 20–100 Kčs.

THEATRE & OPERA

National Theatre (Národní divadlo), Prague 1, Národní třída 2. Tel: 205 364. Opera, ballet, theatre.

Nová Scéna, Prague 1, Národní třída 4. Tel: 206 260. Theatre, opera, Laterna Magika.

Smetana Theatre (Smetanovo divadlo) Prague 1, Wilsonova 8. Tel: 269 746. Opera, ballet.

Laterna Magika, Prague 1, Národní třída 4. Tel: 206 260. A mixture of music, dance and theatre.

Estates Theatre (Stavovské divadlo), Prague 1, Ovocný trh 6. Tel: 227 281. Opera, theatre.

Theatre on the Balcony (Divadlo na Zábradlí), Prague 1, Anenské nám. 5. Tel: 236 0449. Theatre, mime.

Mime Theatre (Branické divadlo pantomimy), Prague 4, Branická 63. Tel: 460 307. Mime.

Spejbl and Hurvinek Theatre (Divadlo Spejbla a Hurvínka), Prague 2, Římská 45. Tel: 251 666. Puppet theatre.

Studio Gag Boris Hybner, Prague 1, Národní třída 25. Tel: 265 436. Mime.

Italian Theatre (Divadlo u Italů), Prague l, Šporkova 13. Tel: 535 181. Cabaret.

Prague Intimate Opera (Komorní Opera Praha), Opera Mozart "A" scéna, Prague 1, Novotného lávka 1. Tel: 265 371. Modern opera.

Supraclub, Prague 1, Opletalova 5. Tel: 224 537.

Music Theatre in Karlín, (Hudební divadlo v Karlině), Prague 8, Křížíkova 10. Tel: 220 895.

Since July 1992, Čedok has been presenting a variation of "Laterna Magika" with the programme "Laterna Animata". This Audio-Video Show, combined with theatre and mime, is based on themes from "Faust and Margaret". Performances are held in the Spirala Theatre in the Prague exhibition centre.

CONCERT HALLS

House of Artists (Dům umělců), Prague 1, Nám. Jana Palacha. Tel: 231 9164. (being renovated)

Municipal House, Smetana Hall (Obecní dům, Smetanova síň), Prague 1, Nám. Republiky 5. Tel: 232 5858.

Atrium, Prague 3, Čajkovského 12. Tel: 274 080.

Palace of Culture (Palác kultury), Prague 4, 5. května 65. Tel: 417 2741.

Janáček Hall (Janáčkova síň), Prague 1, Besední 3. Tel: 530 546.

Mánes Hall in St Agnes Monastery (Mánesova síň kl. sv. Anežky České), Prague 1, U milosrdnych 17. Tel: 231 4251.

Supraclub, Prague 1, Opletalova 5. Tel: 224 537.

Music Hall (Hudební síň Klárov), Prague 1, Klárov 3.

House of the Stone Bell (Dům U kamenného zvonu), Prague 1, Staroměstské nám. 13. Tel: 231 0272.

Foerstrova Hall, Prague 1, Pštrossova 17.

Hall of Mirrors (Zrcadlová síň Klementina), Prague 1, Klementinum 190. Tel: 266 541.

Strahov Monastery (Strahovsky Klášter), Prague 1, Strahovské nádvoří 1. Tel: 538 841.

Emmaus Monastery (Opatství Emauzy), Prague 2, Vyšehradská 49.

New Provost's House (Nové proboštství), Prague 2, K rotundě 10.
House of the Lords of Kunštát and Poděbrady (Dům pánu z Kunštátu a Poděbrad), Prague 1, Řetězová 3.
Schwarzenberg Palais (Schwarzenberský palác), Prague 1, Hradcanské nám. 2.
Lobkovic Palais (Lobkovický palác), Prague 1, Pražský hrad. Tel: 537 306.
A. Dvořák Museum (Muzeum A. Dvořáka), Prague 2, Ke Karlovu 20. Tel: 298 214.
Villa Bertramka, Prague 5, Mozartova 169. Tel: 543 893.
Martinic Palais (Martinický palác), Prague 1, Hradčanskě nám. 8.
Knights Hall in the Waldstein Palace (Rytířský sál Valdštejnského paláce), Prague 1, Valdštejnské nám.
Troja Castle (Trojský zámek), Prague 7. Tel: 845 133.
Kaunitz Palais (Kauniců̊v palác), Prague 1, Panská 7.
Spanish Hall (Spanělsky sál), Prague 1, Pražský hrad (Prague Castle).

MUSIC FESTIVALS

In May/June there is the International Prague Spring Festival. Concerts are given in historical rooms and churches, for example: St Vitus' Cathedral, Hradčany; St James' Church, Jakubská ul.; Church of St Nicholas, Malostranské nám.

In cultural centres such as the Malostranská beseda there are daily musical and artistic performances, ranging from concerts of chamber music to jazz and rock.

CINEMA

While during the last decades Czech film achieved world renown and the Prague public poured into every première, since the collapse of Communism the cinemas have been invaded by Hollywood productions, often in English. Dubbed films are marked with a white square on the posters. The brochure "cinema" put out by PIS every week contains the cinema programme.

A visit to the **Barandov Studios** is a must for all film freaks. They became world famous especially for their children and youth productions such as *Pan Tau*. Today primarily western sponsored films are made here.

There are two cinemas with a difference: the British Cultural Section (Jungmannova 30) shows only English-language films; the Ponrepo (Veletržní 61, tel: 37 92 78) shows black-and-white films in the original languages.

ART COLLECTIONS

There is a vast number of museums whose collections and treasures have already been described in their relevant places in this book. Here is a summary:

NATIONAL GALLERY

The pride of Prague's museums is the **Národní galerie** (National Gallery). It contains seven collections housed in different buildings.

The Sternberg-Palais, Prague 1, Hradčanské nám. (Open: Tuesday–Sunday 10am–6pm) houses the collections of **Classical European Art and French 19th- and 20th-Century Art**. The rooms of the former palace contain such incomparable masterpieces as the *Rosenkranzfeier* by Albrecht Dürer, fragments of an altarpiece by Lucas Cranach, and the *Martyrdom of St Florian* by Albrecht Altdorfer.

Among the works by Italian artists, we find *David with the head of Goliath* and the *St Hieronymus* by Tintoretto, the *Portrait of a Patrician* by Tiepolo or the *View of London* by Canaletto.

The Dutch Collection includes the *Hay Harvest* by Pieter Brueghel the Elder, the *Winter Landscape* by Pieter Brueghel the Younger, the *Martyrdom of St Thomas* by Peter Paul Rubens and the *Rabbi* by Rembrandt.

Amongst the French 19th- and 20th-century contingent are Delacroix, Renoir, *The Green Cornfield* by van Gogh as well as works by Rousseau, Cézanne and Paul Gauguin. The collections with works by Chagall and Picasso are also very popular.

The third collection is to be found in St George's Monastery in Hradčany; Jiřský klášter, Prague 1, Hradčany. Open: Tuesday–Sunday 10am–6pm. It contains **Old Czechoslovakian Art**, including works by the artists Karel Škréta and Jan Kupecký.

The **Modern Art Collection** is to be found in the Městská knihovna (City Library) in Prague 1, Staré Město, nám. primátora dr. V. Vacka 1. Open: Tuesday–Sunday 10am–6pm.

The **Prints and Drawings Collection** in the Palais Kinsky displays Czech, Slovakian and foreign graphics from the last five centuries: Palác Kinských, Prague 1, Staroměstské nám. 12 .

The collection **Czech 19th-and 20th-century Sculpture** is housed outside Prague in Zbraslav Castle. Open: Tuesday–Sunday 10am–6pm (April-November).

The restored St Agnes Monastery houses the collection **Czech 19th-Century Art**, including works by the brothers Quido and Josef Mánes, Karel Purkyně and Mikoláš Aleš: Anežsky klášter, Prague 1, U milosrdných 17.

EXHIBITION HALLS

Royal Palace Belvedere, Prague 1, Chotkovy sady. Open: Tuesday–Sunday 10am–6pm.
Waldstein Palace Riding School, Prague 1, Valdštejnská 2. Open: Tuesday–Sunday 10am–6pm.
Prague Castle Riding School, Prague 1, Hradčany.
Gallery of the Capital Prague: Old Town Hall Cloister, Prague 1, Staroměstské nám. Open: daily 9am–5pm; Old Town Hall, 2. floor exhibition hall. Open: Tuesday–Sunday 10am–5pm.

The Rudolfinian Collection. In the 16th century, the collection of Rudolf II was one of the most important in Europe. Through looting after the Battle of the White Mountain, the annexation by the Habsburgs and Swedes in the Thirty Years' War and subsequent auctioning off, the collection dwindled drastically. Paintings by Rubens and Tintoretto were found when the rooms of the castle were thoroughly searched in the 1960s. Today, the collection has been partially restored and can be seen in the castle gallery. It contains pictures such as the *Scourging of Christ* by Tintoretto and the *Meeting of the Olympic Gods* by Peter Paul Rubens: Rudolfinian Collection, Prague 1, Hradčany (II. courtyard). Open: Tuesday–Sunday 10am–6pm.

FURTHER EXHIBITION HALLS

Mánes House of Fine Arts, Prague 1, Masarykovo nabř. 250; New Hall, Prague 1, Voršilská 3; Václav Spála Gallery, Prague 1, Národní třída 30; U Řečických Gallery, Prague 1, Vodičkova 10; Jaroslav Frágner Gallery, Prague 1, Betlémské nám. 5; Brothers Čapek Gallery, Prague 2, Jugoslávská 20; Gallery D, Prague 5, Matoušova 9; Vincenc Kramář Gallery, Prague 6, Čs. armady 24; ÚLUV Gallery, Prague 1, Národní třída 38; Clementinum Mirrored Hall, Prague 1, Clementinum 190; Melantrich Little Gallery, Prague 1, Jilská 14; Atrium, Prague 3, Čajkovského 1; Palais of the Golden Melon, Prague 1, Michalská 12; Odeon Exhibition Hall, Prague 1, Celetná 11; Clam-Gallas Palais, Prague 1, Husova 20; UBOK, Prague 1, Na příkopě 25–27.

GALLERIES

Hollar Gallery, Prague 1, Smetanovo nábř. 8; Pi-Pi Art Gallery, Prague 1, Národní třída 9–11; GGG Gallery, Prague 1, Husova 10; Klubu pratel Gallery, vytvarného umění U sv. Martina, Prague 1, Uhelný trh; A + G FLORA, Prague 3, Přemyslovská 29; Athena Gallery, Prague 1, U starého hřbitova 46; Forum, Prague 1, Ul. 28. října 12; C-ART Gallery, Prague 1, Loretánské nám. 2; České grafiky Centre, Prague 1, Husova 12; CRUX, Prague 1, Kostel Panny Marie Sněžné, Jungmannovo nám.; Golden Cup Gallery, Prague 1, Nerudova ul.; Carolinum, Prague 1, Ovocný trh 10–18; Letna Gallery, Prague 7, M. Horakové 22; Spektrum, Prague 1, Karlova 4; Gallery 33 Bergman, Prague 2, Vinohradská 79; U Hybernů, Prague 1, Nám. Republiky 4; U sv. Jindřicha Gallery, Prague 1, Jindřišská ul.; Vyšehrad Gallery, Prague 2, K rotundě 10; Chodovská vodní tvrz, Prague 4, Ledvinova 9; VIA ART Gallery, Prague 2, Resslova ul.; Alexy Gallery (Dům slovenské kultury), Prague 1, Purkyňova 4; ART Gallery, Prague 1, Haštalská 10; Gallery of the Association of Czech Photographers, Prague 1, Kamzíkova 8.

Dilo Galleries (selling galleries): Na Újezdě, Újezd 19; Můstek, 28. října 16; Centrum, 28. října 6; Karolina, Železná 6; Platýz, Národní 37; Zlatá lilie, Prague 1, Malé nam. 12; Zlatá ulička, Prague Castle.

MUSEUMS

Apart from the large and small galleries where art treasures from many centuries are on view, Prague has a number of other interesting museums to offer. Note: Museums are closed on Mondays and the days after national holidays. The museums cover a range of topics from natural science to technical and historical. Here is a selection:

National Museum: Národní Muzeum, Prague 1, Václavské nám. The Národní museum on Wenceslas Square is predominantly a natural science museum and contains an extensive mineral collection. In the foyer there are also statues from the Libuše legend by Ludwig Schwanthaler. The Museum Library is also extensive: it contains over a million books. Open: Monday and Friday 9am–4pm; Wednesday, Thursday and Saturday 9am–5pm. Closed: Tuesday.

National Technical Museum: Národní technické muzeum, Prague 7, Kostelní 42. For those interested in technical equipment, measuring equipment, and the first Czech car, the Koprivnitz "President" from 1897, the Národni technické muzeum is an absolute must. There are automobile and locomotive exhibits as well as a photographic exhibition; the Astronomy Department contains sextants from the 16th century, with which Kepler once worked. Open: Tuesday–Sunday 9am–5pm; Museum Archives Wednesday and Thursday 9am–5pm; Museum Library Monday–Friday 9am–4pm.

Historical Military Museum: Vojenské historické muzeum, Schwarzenberský palác, Prague 1, Hradčanské nám. Weapons of all kinds are to be found in the two military museums. In the Historical Military Museum in the Schwarzenberg Palais, many unusual weapons, uniforms and other military hardware are displayed. The collection is one of the largest of its kind in Europe. Open: May–October, Monday–Friday 9am–3.30pm, Saturday and Sunday 9am–5pm.

Military Museum: Vojenské muzeum, Prague 3, U Památníku 2. The histrory of the Czechoslovakian army, and the battles of World War I and II as well as the resistance of the partisans are displayed. Open: Tuesday–Sunday 9.30am–4.30pm.

Náprstek Museum: Náprstkovo muzeum, Prague 1, Betlémské nám. 1. Ethnographic exhibits as well as technical gadgets from the respective countries are displayed in the Museum of Asiatic, African and American Cultures which dates back to the private ethnology museum established by Vojta Náprstek in 1862. Open: Tuesday–Sunday 9am–5pm, Thursday 9am–6pm.

Folklore Museum: Národopisné muzeum, Prague 5, Petrínské sady 98. On the lower section of the Petřín Hill is the former palace of the noble Kinsky family, which today contains the Folklore Museum. The collections include beautiful pottery, glass and toys as well as costumes and furniture from old farmhouses. Especially fine is the belfry from Wallachia which stands next to the museum, as well as the little Orthodox wooden church dating from the 18th century, brought here from the West Ukrainian village of Mukacevo in 1929. Open: Tuesday–Sunday 9am–6pm.

Arts and Crafts Museum: Umĕleckoprůmyslové muzeum, Prague 1, Ulice 17. listopadu 2 (opposite the former Rudolfinum, today's "Artists' House"). Bohemian glass of various periods is displayed in the Arts and Crafts Museum. The glass collection is probably the largest in the world and the library with 100,000 specialist volumes is open to the public. Open: Tuesday–Sunday 10am–5pm.

Post Museum: Poštovní muzeum, Prague 5, Holečkova 10. Philatelists and friends of everything to do with postal history will be in their element in the Post Museum, with its large collection of European stamps. Open: Monday–Friday 8am–2pm, Saturday and Sunday only by appointment.

Museum of the City of Prague: Muzeum hlavního mĕsta Prahy, Prague 8, Nové sady J. Švermy. The Museum of the City of Prague is located at the Sokolovská Metro station. Apart from the countless exhibits describing the history of the city there is also a rare collection dealing with the guilds. But the main attraction remains the famous model of the city constructed between 1826–1834. Accurate to the last detail, it enables a good comparison to be made between the Prague of 160 years ago and the Prague of today. Open: Tuesday–Sunday 10am–5pm.

National Literature Museum: Památník národního písemnictví, Prague 1, Strahovské nadvoří 132. The Strahov Gospels from the 9th century consist of 218 handwritten manuscripts which appear to have come from the monastery workshop in Trier. These gospels and a host of other interesting exhibits can be viewed in the Theologians' and Philosophers' Halls of the Monastery Library. The adjacent rooms of the monastery contain the Museum of Czech Literature, with a collection of around 50,000 objects including a letter from the hand of Jan Hus. Also worth seeing is the reconstruction of a 17th-century press. Open: Tuesday–Sunday 9am–5pm.

Museum of Musical Instruments: Muzeum hudebních nástrojů, Prague 1, Lázeňská 2. Prague is not only a city of architecture and literature, but also of music. This fact is mirrored in the museums. The music lover has a total of four important places to go in Prague, and he should really try to see all of them. The Collection of Old Musical Instruments is the second largest of its kind in the world. Apart from old musical instruments, there is also a large collection of musical scores from various archives. Open: Saturday and Sunday 10am–noon, 2–5pm; weekday visits subject to arrangement.

Antonín Dvořák Museum: Muzeum Antonína Dvořáka, Prague 2, Ke Karlovu 20. The term "Villa Amerika" dates from the 19th century and is so named after a pub. This beautiful little building, designed by Kilian Ignaz Dientzenhofer and built between 1717–1720 for Count Michna, today contains the Antonín Dvořák Museum. The rooms contain manuscripts, documents, photographs and letters to such important personalities as Johannes Brahms or the conductor Bülow. Open: Tuesday–Sunday 10am–5pm.

Bedřich Smetana Museum: Muzeum Bedřicha Smetany, Prague 1, Novotného lávka 1. The Smetana Museum is housed in the former waterworks on the banks of the Vltava. Manuscipts, notes and other documents illuminate the life and work of this great Czech composer. Open: Monday–Sunday 10am–5pm. Closed: Tuesday.

Museum of Czech Music: Muzeum české hudby, expozice W. A. Mozarta, Prague 5, Mozartova 15. Music lovers know that Prague is also linked with the name of Wolfgang Amadeus Mozart – catchword *Prague Symphony* or *Don Giovanni*. Mozart lived in Prague in the Villa Bertramka, which today houses the Mozart Museum. The harpsichord and piano are the very same instruments as those on which Mozart composed his music. Most of the furniture also dates from the time of his stay. Open: Monday–Friday 1–3pm, Saturday and Sunday 10am–noon and 1–4pm.

Judaic Museum: Different departments of the National Judaic Museum can be found in the synagogues of the old Prague Ghetto. It is a tragic irony of fate that it was the Nazis themselves, to whom 90 percent of Prague's Jews fell victim, who intended to erect "The Museum of the Extinct Jewish Race", by collecting together cultural and artistic objects from all over the country. They thus laid the foundation for the present museum, which was then founded by the Czech government in 1950. Due to restoration work it will not be possible to see the exhibits in the Pinkas Synagogue for some time.

Nightlife

While during the socialist era, there was no nightlife to speak of in Prague (the pubs all closed at 10pm), the metropolis on the Vltava is now doing its best to be like any other international city. If you're not keen on plunging in on your own, then put your trust in Čedok which runs the "O.K. Revue in Prague" in the **Alhambra Club**. Another alternative is the programme "Dinner and Show". For those in search of "good old Bohemia", the official travel agency offers two entertaining evening events:
Prague Festival/Country Fair from May to October in the Hotel International. One big party with typical Bohemian fare and as much beer as you like; a colourful folklore programme, brass band, dance and lots of fun.
"Bohemian Fantasies" from April to October. A musical odyssey through the history of Czechoslovakia. Held in the hall of the Lucerna Palace which was built by Václav Havel's grandfather, if offers a varied programme, with classical music, folklore and other entertainment.

NIGHT CLUBS

The night clubs offer variety programmes, revues or cabarets and naturally live music to dance to. They are smart places and the cost is correspondingly high. There are an increasing number of striptease shows. A popular address is the Alhambra with its "Nightshow" of music, black theatre and the usual variety numbers: **Alhambra Nightshow**, Prague 1, Václavské nám. 5. Tel: 220 467. Open: 8.30pm–3am.

The **Est-Bar** in the Hotel Esplanade is a very elegant place indeed. The reputation of the hotel equally applies to the night club with its varying programme. Est-Bar, Prague 1, Washingtonova 19. Tel: 222 552. Open: 9pm–3am.

The two nightspots **Jalta Club** and **Jalta Bar** are to be found in the venerable hotel of the same name. There is an orchestra, variety and disco: Jalta Club, Jalta Bar, Prague 1, Václavské nám. 45. Tel: 265 541-9. Open: 9pm–3am.

One of the best programmes in Prague's nighlife scene is offered by the **Interconti Club** in the Inter-Continental Hotel: Interconti Club, Prague 1, Náměstí Curieových. Tel: 28 9. Open: 9pm–4am.

The Lucerna Palace also offers variety entertainment. The **Lucerna Bar** is one of the largest of Prague's bars and dance floors; it is also a venue for concerts: Lucerna Bar, Prague 1, Štěpánská 61. Tel: 235 0888. Open: 8.30pm–3am.
Monica, Prague 1, Charvátova 11. Open: all night. Often live jazz and striptease.
Narcis, Prague 1, Melantrichova 5. The usual bar atmosphere.
The Park Club in the Park Hotel is a popular haunt of business people: Park Club, Prague 7, Veletrzní 20. Open: 8.30pm–3am, Friday and Saturdays until 4am.
The Tatran Bar offers an exciting programme and has a coffeehouse with dancing on a glass floor: Tatran Bar, Prague 1, Václavské nám. 22. Open: 8.30pm–4am, dance-café 5pm–midnight.
Varieté, Prague 1, Vodičkova 30. The famous revue theatre has now become a strip joint.

Here are a few of the coffeehouses with dancing and bars: **Alfa**, Prague 1, Václavské nám. 28. Open: 6pm–1am; **Astra**, Prague 1, Václavské nám. 4. Open: 10am–midnight; **Barbara**, Prague 1, Jungmannovo nám. 14. Open: 8.30pm–4am; **T-Club**, Prague 1, Jungmannovo nám. 14. Open: 8.30pm–4am.

DISCOTHEQUES

Most discotheques are located in the Wenceslas Square area. Here one is likely to meet a very mixed public, and most locals avoid them. The local ladies that you come across might easily be prostitutes. Female tourists who enter such establishments unaccompanied must therefore expect constant advances.

There is usually a bouncer at the door and admission prices vary from between 50–100 Kčs. Jeans and trainers are not accepted as being suitable attire. The botels on the Vltava (*see Where to Stay*) offer an attractive alternative.

Classic Club, Pařížska 4 (1. floor). Meeting place for artists. Open: until 3am.
A Scéna (in the Smetana Museum on the Vltava), Novotného lávka 1. Prague Scene, usually bursting at the seams later on.
Starlight Discoteka, Palác Kultury, Vyšehrad. Nice view of Prague, but the usual disco sound. Open: until 6am at weekends.
New D Club, Vinohradská 38. Mostly locals, few tourists.
Peklo (Hell). The latest rage. Gothic catacomb disco in the Strahov Monastery. Admission for members only.
Adria, Národni třída 40. Open: daily from 8.30pm.
Astra, Václavské náměstí 28. Open: daily 7pm–2am.
Habana, V jámě 8. Open: daily 9.30pm–4.30am.
Rostov, Václavské náměstí 21. Open: daily 8pm–3am.
Video-Disco, in the Hotel Zlatá Husa, Václavské nám. 7. Open: daily 7.30pm–2am.

Botel Admirál, Horejší nábřezí.
Botel Albatros, Nábřezí L. Svobody.
Botel Racek, Dvořecká louka.

JAZZ, ROCK & POP

Prague is not only a city of classical music. There is a rich calendar of rock and pop events, and since the fall of the iron curtain, the city has hosted many an international star band. For friends of jazz, Prague was always an important venue, especially for the traditional kind:

Agharta Jazz centre, Krakovská 5. Swing and new rhythms.

Club 007, Kolej Strahov (student residence), Prague 1, Spartakiadní 7. Tel: 354 441. Student club with ever-changing bands. Open: Saturday from 8pm–midnight.

Lidový dům, Prague 9, Emanuela Klímy 3. Tel: 823 434. Heavy metal and hard rock scene. Open: until 1am.

Malostranská Beseda, Prague 1, Malostranské náměstí 21. Variable programme, partly jazz.

Metro, Prague 1, Národní třída 20. Tel: 262 085. Jazz Club, bands representing all styles. Open: Tuesday–Friday until 2am.

Na Chmelnici, Prague 8, Koněvová 219. Tel: 828 598. First address for local talent.

Palác kultury (Palace of Culture), Prague 4, Ulice 5. května 65. Tel: 417 2741. A multi-purpose 6-floor centre with a variety of concert halls and stages.

Reduta, Prague 1, Národní třída 20. Tel: 203 825. The best known jazz club in the city. All styles, always full. Open: Monday–Friday until 2am.

Jazz Art Club, Prague 2, Vinohradská 40. Tel: 757 654. Modern jazz, Prague bands. Open: daily 9pm–2am, except Monday.

Rock Café, Prague 1, Národní třída 20. Meeting place of the Underground scene.

Sněhobílá kočka, Prague 9. Českomoravská 15. Rock and beer until 4am.

Újezd, Prague 1, Újezd 18, New Wave und Underground. Open: until 6am.

Press Jazz Club, Prague 1, Pařížská 9. New club with a varying programme. Open: Monday–Saturday 9pm–2am.

Viola Wine Bar, Prague 1, Národní třída 7. Tel: 235 87 79. Intimate stage for the literary scene with a programme of jazz on Saturdays. Open: 8pm–midnight.

CASINOS

Casinos seem to be shooting up like mushrooms in Prague. There are now about a dozen clubs where you can gamble your money. But it is seldom the tourists who occupy the Black Jack or Roulette tables, rather a special breed of *nouveau riche* locals, who may have earned their money on the black market and need to launder it. They take the matter very seriously indeed. Only foreigners give tips to the groupiers.

There are four hotels that run casinos: Forum, Ambassador, Palace and the Diplomat Club. Bets are placed in US-Dollars. The casinos are open from 9pm–4am.

SHOPPING

Prague falls sadly short in comparison to other European metropolises when it comes to shopping. Despite this, the Czech capital is considered to be the absolute shopping paradise of what used to be known as the Eastern Block cities. Citizens from Poland and former East Germany used to come here for the weekend, shop like mad during the day and amuse themselves in the evening at nightclubs and pubs. And, as a matter of fact, there are quite a number of small speciality shops and a few well-stocked department stores located in the centre of Prague.

On the whole, Western European visitors seem to be mainly interested in the porcelain and crystal shops. Bohemian glass and china is known and held in high esteem throughout the world on account of its exceptional quality and extraordinarily low price. Due to this, tourists frequently spend hours waiting in line in front of the national retail outlets. Because the china and glass companies can scarcely keep up with the demand for their wares any more, they have taken to selling old stock as the majority of their new products is reserved for export.

Tourists don't seem to be particularly choosy though. Once they've been bitten by the shopping spree bug, they tend to buy just about anything that is known to cost a fair amount more in other industrial countries, but which still can be purchased in Prague for relatively cheap prices. Nowadays however, it's almost impossible to find a good deal in the antique shops. Antique dealers have become wise to the Western predilection for their wares and have altered their prices accordingly. As a compensation, gallery sales are booming. Works by artists who until 1989 were excluded from the circle of artists' associations are now being exhibited and sold. Private, more amateur arts and crafts items are also finding a ready market. Plain and carved wooden objects produced throughout the country are for sale at authorised Czech retail outlets. "Wandering street vendors" – usually young people on the Charles Bridge and in the street Na příkopě

– sell handmade goods, for instance marionettes and costume jewellery. Quite recently open-air markets have sprung up in the pedestrian zones between Náměstí Republik and Wenceslas Square. Clothes, household articles, souvenirs and odds and ends of every description eventually find buyers amongst the people strolling by.

Buying books written in both English and German is another popular pastime. But since the printing of books is no longer subsidised by the government, Czech publishing houses are on the verge of financial ruin and are looking for help from foreign investors. For the people of Prague this means books are getting to the point where practically no one can afford them; for foreign visitors, however, buying a book remains a relatively inexpensive pleasure.

Classical music buffs can look forward to acquiring good quality and fairly inexpensive records in Prague. (CD's and cassettes are still much harder to find.)

If you're looking for something typically Bohemian to bring home with you as a gift, how about a Prague ham? Despite the fact that butcher shops are numerous in the capital, authentic Prague ham is only rarely to be found. This being the case, it may prove less troublesome to procure a bottle of the herb liqueur Becherovka or Slivowitz. Fruity wines from Bohemia and Moravia are also gifts sure to be appreciated.

Visitors preparing to head out for an extensive round of shopping should bear in mind that such an excursion can turn out to be fairly time-consuming. Shopping can really turn into a test of patience, especially on the weekends when both tourists and Prague natives hit the streets. Taking this into account it's a good idea to get your errands and shopping accomplished during the week and to reserve the weekend for visiting the interesting sights that the city has to offer. Whatever you can't find in the centre of Prague, you can nearly be sure that you won't be able to find it in any of the other city districts either. The following list of some of the special retail shops located in the centre should be of help to you in finding what you're looking for.

As a rule most shops are open 10am–6pm Monday–Friday, and 10am–2pm Saturday. (Stores are closed on Sunday.) Department stores remain open somewhat later.

International Bookshops

Kniha, Štěpánská 12 (in the courtyard), Prague 1. Open: daily 9am–6pm, Saturday 9am–1pm. Closed: Sunday.
Zahraniční Literatura, Vodičkova 41 (in the Alfa Arcade, Wenceslas Square entrance), Prague 1. Open: daily 9am–6pm, Saturday 9am–1pm. Closed: Sunday.
Kniha, Na příkopě 27, Prague 1.

Second-Hand Book Shops

AD plus, Prague 1, Václavské nám. 18. Also in Prague 1: Dlážděná 5; Mostecka 22; Karlova 2; Ul. 28. října 13. Prague 2: Ječná 26.

Hat Shops

Tonak, Celetná 30, Prague 1. Open: daily 9am–6pm, Saturday 9am–1pm. Closed: Sunday.

Antiques

Antikvita, Prague 1, Panská 1. Also in Prague 1: Mikulandská 7; Václavské nám. 60; Uhelny trh 6; Můstek 3. Prague 2: Vinohradská 45. Prague 7: Šimáčková 17.

Jewellery

Bijoux de Boheme, Prague 1, Dlouhá. Also in Prague 1: 28. října 15; Národní třída 25; Na příkopě 15; Václavské nám. 53.
Galerie Vlasta Wasserbauerova, Staroměstské náměstí 5, Prague 1. Hand-made individual pieces.

Fashion Jewellery

Na příkopě 12, Prague 1; Staroměstské náměstí 6, Prague 1. Beads, stones.

Garnet Jewellery

Granat, Prague 1, Václavské nám. 28 (Alfa Arcade).

Glass and Porcelain

Bohemia Moser, Na příkopě 12, Prague 1. Especially beautiful pieces.
Bohemia, Pařižská 1, Prague 1.

Crystal, Lamps, Crockery

Krystal, Václavské nám. 30, Prague 1.

Chandeliers

Superlux, Prague 1, Hybernská 32.
Lux, Prague 1, Na příkopě 16.

Objets d'Art

Dilo, Vodičkova 32, Prague 1.

Arts and Crafts

Česká jizba, Karlova 12, Prague 1. **Krásná jizba**, Národní 36, Prague 1. **Slovenská jizba**, Prague 1, Václavské nám. 40. Open: daily 9am–12 noon, 2pm–7pm; Saturday 9am–1pm.
UVA, Na příkopě 25, Prague 1. **A+G Flora**, Přemyslova 29, Prague 3. Tel: 27 17 16. No regular opening times; ring beforehand.

Music and Instruments

Prague 1: Jungmannovo nam. 17 and 30; Na příkopě 24.

Records

Prague 1: Jungmannova 20; Celetná 8; Václavské náměstí 17 and 51; Vodičkova 20.

Sports articles

Dům sportu, Jungmannova 28, Prague 1.
Sportovní potřeby, Vodičkova 30, Prague 1.

Weapons and Hunting

Lověna, Prague 1, Hybernská 3.
Lověna, Prague 1, Národní třída 38.

DEPARTMENT STORES

The department stores with the largest selection of goods (including gift articles, fabric, clothes, dishes, shoes, perfume, drugstore items, groceries, travel accessories, writing materials, electrical goods, books, etc.) are: **Bílá labut'**, Prague 1, Na Porící; **Detský dům** (House of Children), Prague 1, Na příkopě 15; **Družba**, Prague 1, Václavské nám. 21; **Dům elegance** (House of Elegance), Prague 1, Na příkopě 4; **Dům kožešin** (House of Furs), Prague 1, Železná 14; **Dům módy** (House of Fashion), Prague 1, Václavské nám. 58; **Dům obuví** (House of Shoes), Prague 1, Václavské nám.; **Kotva**, Prague 1, Náměstí Republiky 8; **Máj**, Prague 1, Národní třída 26. Most large department stores have a supermarket in the basement. A delikatessen shop to be particularly recommended is: **Dům potravin**, Václavské náměstí 59, Prague 1.

SPORTS

TENNIS

For years now, Czech tennis players, both men and women, have found themselves right at the top of the world rankings. The Czech Tennis School has an international reputation, and Steffi Graf certainly wouldn't have got where she is were it not for her trainer Pavel Složil.

The Prague Tennis Arena on the island of Štvanice in the Vltava, the venue for Davis Cup and Grand Prix tournaments, is considered to be one of the most beautiful such complexes in Europe. But tourists can also play tennis in Prague. All you have to do is book with Čedok, which between June and September organises courses for both beginners and advanced under the instruction of top trainers.

The best known tennis courts are at:

Štvanice Stadium Prague 7. Tel: 2316323. About £7 per hour.

Also indoor courts at Štvanice, as well as at: Prague 7, Kostelní. Tel: 37 36 83, and Prague 7, Stromovka. Tel: 325479, 324850. An all-year-round tennis school is offered, for example, by the **Club Hotel** Praha Průhonice, Průhonice near Prague. Tel: 00422-723241-9.

RIDING

Pony treks take place in Konopiště near Prague and can be booked through Čedok.

The Prague race course (Státní závodiště) is situated outside the city in the southern suburb of Chuchle.

ICE SKATING

Slavia Praha IPS Stadium, Prague 10, Vrsovice.
Štvcanice sports complex, Prague 7, under the Hlávka Bridge.

SWIMMING

The open-air pools on the Vltava are not recommended on account of the poor water quality. The modern ☆☆☆☆☆-star hotels such as the Forum and Panorama have swimming pools and saunas usually open to non-guests as well (about 200 Kčs for 2 hours, open 9am–8pm).

Podolí Stadium, Prague 4, Podolská 74. Tel: 427384. Open-air pool, 26° C, with sauna and steam bath. Open: weekdays 6am–10pm, Saturday and Sunday 8am–10pm.

Indoor pools:
Dům kultury Klárov, Prague 1, Nábřeží kapit. Jařose.
Swimming pool in the recreation park, Prague 7, Stromovka.Tel: 375404.

GOLF

Prague also has an 18-hole golf course, adjacent to the Hotel Golf Praha in the district of Motol in the west of the city:
Hotel Golf, Prague 5 (Motol), Plzeňská. Tel: 59 66 93. Open: daily 8am–8pm (1 April–30 October) and from 9am to dusk in the off-season.

LANGUAGE

USEFUL VOCABULARY

The menus in the top class and hotel restaurants are usually written in two or more languages. And pretty well every waiter in Prague speaks either German or English. Nevertheless, here is a short list of most of the food and drink you're likely to encounter.

BEVERAGES

Becherovka	bitter cordial
čaj	tea
černy čaj	black tea
káva	coffee
černá káva	turkish coffee
káva s mlékem	milk coffee
videnská káva	coffee with whipped cream
limonáda	lemonade
pivo	beer
malé pivo	small beer
černé pivo	dark beer
svetlé pivo	lager
točené pivo	draught beer
slivovice	slivovitz
víno	wine
bilé víno	white wine
červené víno	red wine
voda	water

FOOD

bažant	pheasant
biftek	steak
bramborák	potato fritter
brambory	potatoes
buchty	Bohemian sweet dumplings
chléb	bread
bílý chléb	white bread
černy chléb	brown bread
cukr	sugar
drůbež	poultry
fazole	beans
guláš	goulash
houby	mushrooms
hovězí	beef
hovězí pečené	roast beef
hovězí vařené	boiled beef
hruška	pear
husa	goose
houska	roll
jablka	apple
kachna	duck
kančí	wild boar
kapr pecený	fried carp
kapr smažený	garnished carp
kapr varený	boiled carp
kapusta	savoy cabbage
kaše bramborová	mashed potatoes
knedlíky bramborové	potato dumpling
knedlíky houskové	white bread dumpling
knedlíky ovocné	fruit dumpling
králík	rabbit
krocan	turkey
kuře	chicken
kuře smažené	roast chicken
kyselé zelí	sauerkraut
ledvinky	kidneys
máslo	butter
meruňky	apricots
mrkev	carrots
ořechy	nuts
ovoce	fruit
palačinky	thin pancakes
párky	sausages
pečené	roast meat
pečivo	biscuits
polévka	soup
polévka dršťková	tripe soup
pstruh	trout
rajčata	tomato
rostěnka	roast meat
ryba	fish
rýže	rice
salám	sausage, salami
salát	salad
sardinký	sardines
sekaná	meat loaf
sladký	sweet
slaný	salty
srncí	roast venison
štika	pike
šunka	ham
telecí	veal
topinky	toast
třešně	cherries
uzenina	smoked meat
vejce do skla	egg in a glass
vejce na měkko	soft boiled egg
vejce na tvrdo	hard boiled egg
vepřová	roast pork
zajíc	hare
zelenina	vegetables
zmrzlina	icecream
zveřina	game

OTHER USEFUL WORDS

kavárna	coffee house
restaurace	restaurant

hostinec	pub
vinárna	wine bar
snídaně	breakfast
oběd	lunch
večeře	supper
volno	free
obsazeno	occupied
stůl	table
židle	chair
nůž	knife
vidlička	fork
lžice	spoon
talíř	plate
sklenice	glass
číšník	waiter
vrchní	head waiter
sevírka	waitress
ubrousek	serviette
jídelní lístek	menu
specialita	speciality
párátko	toothpick

LANGUAGE NOTES

In the Czech language stress is always given to the first syllable. Long vowels are indicated by an accent (the long u by an accent or small circle):

á, é ,í, ó, ú, ů, ý

l and r can be pronounced as half vowels:

Plzeň = Pilsen

ý long "e".
ou pronounced "show"
ě pronounced yea
č pronounced "church"

Consonants:

ř prononounced with a silibant
š pronounced sh
ž pronounced with a "j" as in "journey"

SPECIAL INFORMATION

TOURIST INFORMATION

The Prague Information Service (PIS) has an office in Prague 1, Na příkopě 20. Here you can obtain all information about Prague, including city maps and the magazine *One Month in Prague*, containing the most important information and addresses. There is a monthly programme published in English and German with all the theatre and concert events as well as tips for exhibitions and a selection of important cultural events. PIS also maintains an office in the Hradčanská Metro station (Line A). For those who can read Czech, there is the more detailed *Přehled kulturních pořadu v Praze*, which lists up events district by district.

Travel information and bookings are generally organised through Čedok in Prague 1, Na příkopě 18. Tel: 212 7111. Tourist programmes, cultural programmes, handling of tickets, Prague 1, Bílkova 6 (Inter-Continental Hotel). Tel: 231 8855.

In addition to Čedok, tickets for sporting events can be purchased from the central advanced booking office in Prague 1, Spálená 23.

TRAVEL AGENCIES

This business is booming like no other. Here is a list of the main agencies, although new private enterprises are being established all the time:

Autoturist, Prague 2, Na rybníčku 16. Tel: 203 3558; Prague 10, Limuzská 12. Tel: 773 455.

AVE Ud., Prague 2, Sokolská 56. Tel: 205 229; Prague 2, Main Station, Wilsonova 8. Tel: 236 2560, 236 3075.

Balnea, Prague 1, Pařížská 1. Tel: 232 3767, 292 868. Specialises in the spa resorts.

Bohemia Tour, Prague 1, Zlatnická 7.

CKM, Prague 2, Žitná 12. Tel: 299949. For young people.

ČSD, Prague 1, Na příkopě 31. Tel: 236 3238; Prague 2, Wilson Station (Main Station).
Tel: 235 2884, Fax 235 2752; Prague 7, Holesovice Station.

Pragotur, Prague 1, U Obecního domu.
Tel: 231 728.

Rekrea, Prague 1, Revoluční 13. Tel: 231 0633. Room reservations.

Sport-Turist, Prague 1, Národní třída 33. Tel: 263 886. Air and train tickets.

TRANSLATION SERVICES

Artlingua, Prague 2, Myslíkova 6. Tel: 295 169, 295 597, 293 741, 294 198. Interpreting and translation services offered in 30 languages; organisation of congresses and press conferences.

Kahlen Service, Prague 4, Strakonicka 510. Tel: 345 346.

Babel Service, Prague 4, Palác Kultury (Palace of Culture). Tel: 692 6741.

TAP, Prague 1, Helichova 1. Tel: 547 844.

USEFUL ADDRESSES

CONSULATES & EMBASSIES

Algeria, Prague 6, Korejská 16. Tel: 312 0758.

Argentina, Prague 1, Washingtonova 25. Tel: 22 854.

Austria, Prague 5, Victora Huga 10. Tel: 546 557.

Belgium, Prague 1, Valdštejnská 6. Tel: 534 051.

Brasil, Prague 1, Bolzanova 5. Tel: 229 254.

Canada, Prague 6, Mickiewiczova 6. Tel: 312 0251.

CIS, Prague 6, Pod kaštany 16. Tel: 381 940.

Denmark, Prague 2, U Havlíčkových sadů 1. Tel: 254 715.

Finland, Prague 2, Dřevná 2. Tel: 205 541.

France, Prague 1, Velkopřevorské nám. 2. Tel: 533 041.

Germany, Prague 1, Vlašská 19. Tel: 532 351.

Greece, Prague 6, Na Ořechovce 19. Tel: 354 279, 356 723.

Hungary, Prague 6, Mičurinova 1. Tel. 365 041.

India, Prague 1, Valdštejnská 6. Tel: 532 642.

Italy, Prague 1, Nerudova 20. Tel: 530 666.

Japan, Prague 1, Maltézské nám. 6. Tel: 535 751.

Mexico, Prague 7, Nad Kazankou 8. Tel: 855 1539.

Morocco, Prague 6, K Starému Bubenči 4. Tel: 329 4404.

Netherlands, Prague 1, Maltézské nám. 1. Tel: 531 378.

Norway, Prague 6, Na Ořechovce 69. Tel: 354 56.

Poland, Prague 1, Valdštejnska 8. Tel: 536 951.

Portugal, Prague 7, Bubenská 3. Tel: 878 472.

Romania, Prague 1, Nerudova 5. Tel. 533 059.

Spain, Prague 6, Pevnostní 9. Tel: 327 124.

Sweden, Prague l, Úvoz 13. Tel: 533 344, 533 865.

Switzerland, Prague 6, Pevnostní 7. Tel: 328 319.

United Kingdom, Prague 1, Thunovská 14. Tel: 533 347, 533 370.

USA, Prague 1, Trziste 15. Tel: 53 66 41.

EMBASSIES ABROAD

Argentina, Embajada de la Republica Federativa Checa y Eslovaca, Av. Figueroa Alcorta 3240, Buenos Aires.

Austria, Botschaft der Tschechischen und Slowakischen Föderativen Republik, Penzinger Strasse 11-13, 1140 Wien.

Australia, Embassy of the Czech and Slovak Federal Republic, 47 Culgoa Circuit, O'Malley, Canberra ACT 2606.

Belgium, Ambassade de la Republique Federative Tcheque et Slovaque, 152 Avenue A. Buyl, 1050 Bruxelles.

Brazil, Embaixada da Republica Federativa Tcheca y Eslovaca, Avenida das Nacoes, lote 21, Caixa Postal 07-0970, Brasilia DF.

Bulgaria, Embassy of the Czech and Slovak Federal Republic, Bulvar Janko Sakazov 9, Sophia.

Canada, Embassy of the Czech and Slovak Federal Republic, 50 Rideau Terrace, Ottawa, Ontario, K1M 2A1.

Cyprus, Embassy of the Czech and Slovak Federal Republic, 7, Kastorias Street, PO Box 1165, Nicosia.

Denmark, Embassy of the Czech and Slovak Federal Republic, Ryvangs Alle 14-16, 2100 Kobenhavn O.

Finland, Embassy of the Czech and Slovak Federal Republic, Armfeltintie 14, 00150 Helsinki 15.

France, Ambassade de la Republique Federative Tcheque et Slovaque, 15, Avenue Charles Floquet 75 007 Paris.

Germany, Botschaft der Tschechischen und Slowakischen Föderativen Republik, 5300 Bonn, Ferdinandstrasse 27.

Greece, Ambassade de la Republique Federative Tcheque et Slovaque, 6, Rue Seferis, Palaio Psychico, 15452 Athenes.

Iceland, Embassy of the Czech and Slovak Federal Republic, Smargata 16, 101 Reykjavik.

India, Embassy of the Czech and Slovak Federal Republic, 50/M Niti Marg, Chanakyapuri, New Delhi – 110021.

Ireland, Trade Mission of the Czech and Slovak Federal Republic, Confederation House of Irish Industry, Kildare Street, Dublin 2.

Israel, Embassy of the Czech and Slovak Federal Republic, Zaitlin Str. 23, PO Box 16361, 61664 Tel Aviv.

Italy, Ambasciata della Repubblica Federativa Ceca e Slovacca, Pontificio Collegio, Nepomuceno, Via Monte Santo 25, 00195 Roma.

Japan, Embassy of the Czech and Slovak Federal Republic, 16-14, Hiroo 2-chome, Shibuya-ku, Tokyo 150.

Malaysia, Embassy of the Czech and Slovak Federal Republic, 32 Jalan Mesra (off Jalan Damai), PO Box 12496, 50780 Kuala Lumpur.

Mexico, Embajada de la Republica Federativa Checa y Eslovaca, Cuvier, No. 22, Colonia Nuevo Anzures, Mexico 5 DF.

Netherlands, Ambassade de la Republique Federative Tcheque et Slovaque, Parkweg 1, 2585 JG Den Haag.
Norway, Embassy of the Czech and Slovak Federal Republic, Thomas Heftyes gate 32, 0264 Oslo 2.
New Zealand, Embassy of the Czech and Slovak Federal Republic, 12 Anne Street, Wadestown, PO Box 2843, Wellington.
Poland, Ambassada Czeskiej i Slowackiej Republiki Federacyjnej, Koszykowa 18, Warszawa, Skr. poczt. 00-555.
Portugal, Embaixada da Republica Federativa Checa e Eslovaca, Rua Pinheiro Chagas No. 6, Lisboa.
Romania, Ambassade de la Republique Federative Tcheque et Slovaque, Strada Ion Ghica 11, Sector 4, Bucuresti.
Russia, Embassy of the Czech and Slovak Federal Republic, Moscow D-47, ul. J. Fucika 12/14.
South Africa, Consulate General of the Czech and Slovak Federal Republic, LO 3, Matroosberg Road, Waterkloof Park, Pretoria, PO Box 3326.
Spain, Embajada de la Republica Federativa Checa y Eslovaca, Pinar 20, 28006, Madrid 6.
Sweden, Embassy of the Czech and Slovak Federal Republic, Floragatan 13, 11431 Stockholm.
Switzerland, Ambassade de la Republique Federative Tcheque et Slovaque, Muristrasse 53, 3000 Bern 16.
United Arab Emirates, Embassy of the Czech and Slovak Federal Republic, PO Box 27009, Abu Dhabi.
United Kingdom, Embassy of the Czech and Slovak Federal Republic, 25 Kensington Palace Gardens, London W8 4QY.
United States of America, Embassy of the Czech and Slovak Federal Republic, 3900 Linnean Avenue NW, Washington DC, 20008.

CEDOK REPRESENTATIVES

Austria, Parkring 12, 1010 Wien. Tel: 43/222 512 1374, Fax: 43/222 512 591 685.
France, Čedok France S.a.r.l., Avenue de L'Opera 32, 75002 Paris. Tel: 33/1 4742 7487, Fax: 33/1 4924 9946.
Germany, Čedok Reisen GmbH, Kaiserstrasse 54, 6000 Frankfurt 1. Tel: 49/69 274 017, Fax: 0049/69 235 890.
Italy, Čedok Italia S.R.L., Via Piemonte 32, 00 187 Roma. Tel: 39/6 483 406, Fax: 39/6 482 8397.
Sweden, Nordisk – Čedok, Sveavaegen 9–11, 111 57 Stockholm. Tel: 46/8 207 250, Fax: 46/8 200 090.
Switzerland, Čedok – Tschechoslowakisches Reisebüro GmbH, Uraniastrasse 34/2, 8025 Zürich. Tel: 41/1 211 4245, Fax: 41/1 211 4246.
United Kingdom, Čedok London Limited, Czechoslovak Travel Bureau, 17–18 Old Bond Street, London W1X 4RB. Tel: 44/71 629 6058, Fax: 44/71 493 7841.
USA, Cědok, Czechoslovak Travel Bureau Inc., 10 East 40th Street, New York N.Y. 10016, 1/212 609 9720, Fax 1/212 418 0597.

FURTHER READING

Kafka, Franz, *America/The Trail/The Castle*. Translated by W. and E. Muir, Penguin Modern Classics.

Kafka, Franz, *Description of a Struggle and other stories*. Translated by W. and E. Muir, Penguin Modern Classics.

Kafka, Franz, *Diaries*. Translated by W. and E. Muir, Penguin Modern Classics.

Hašek, Jaroslav, *The Good Soldier Schweik*. Translated by Sir C. Parrot, Heinemann.

Vaculík, Ludvík, *A cup of Coffee with my Interrogator*. Translated by G. Theiner, Readers International.

Vaculík, Ludvík, *Prague Chronicles*. Translated by G. Theiner, Readers International.

Seifert, Jaroslav, *Selected Poetry*. Translated by E. Osers, Andre Deuts.

Neruda, Jan, *Tales of the Little Quarter*. Translated by E. Pargetwr, Greenwood Press London.

Art/Photo Credits

Photography by

Page 27, 30, 32, 34/35, 36, 37, 39, 40. 41, 48, 49, 51, 52, 53, 128, 129, 170, 181, 188, 191, 192	**Archives for Art & History**
8/9, 10/11, 22, 24, 28, 31, 33, 38, 58/59, 60/61, 62/63, 65, 66/67, 70/71, 76/77, 81, 82, 84, 85, 86, 87, 88, 89, 90, 92/93, 95, 96/97, 99, 100, 102, 104L, 104R, 105, 107, 108, 109, 114/115, 116, 119, 120, 121, 122, 123, 125, 126, 127, 136, 139, 140, 141, 144/145, 148, 149, 150, 151, 153, 154, 157, 158/159, 1690, 161, 162, 163, 168/169, 171, 173, 174/175, 178, 180, 182, 184, 200/201, 208, 209, 210	**Bodo Bondzio**
12/13, 14/15, 54/55, 57, 64, 68/69, 72, 78, 80, 91, 106, 111, 113, 124, 130/131, 132, 134, 135L, 135R, 137, 138L, 138R, 146, 147, 152, 155, 164, 165, 176, 177, 179, 183, 185, 186/187, 190, 197, 198, 199, 202, 204, 205, 206/207	**Joachim Chwaszcza**
43, 44, 45, 46/47, 50, 98, 112	**Collection Pavel Scheuffler**
56	**Fotografia Panska 4**
118, 142/143	**Jens Schumman**
194	**Sven Simon**
Maps	**Berndtson & Berndtson**
Illustrations	**Klaus Geisler**
Visual Consultant	**V. Barl**

INDEX

A

B

C

D

E – F

G

H

S

T

U – V

W

Y – Z

A
B
D
E
F
G
H
I
J
a
b
c
d
e
f
g
h
i
j
k